The Golden Anniversary Celebration of the National Association of Mathematicians

CONTEMPORARY MATHEMATICS

759

The Golden Anniversary Celebration of the National Association of Mathematicians

AMS Special Session
The Mathematics of Historically Black Colleges
and Universities (HBCUs) in the Mid-Atlantic
January 17, 2019 | Baltimore, Maryland

MAA Invited Paper Session
The Past 50 Years of African Americans in the Mathematical Sciences
January 18, 2019 | Baltimore, Maryland

Haynes-Granville-Browne Session of Presentations by Recent Doctoral Recipients
January 18, 2019 | Baltimore, Maryland

2019 Claytor-Woodard Lecture
January 19, 2019 | Baltimore, Maryland

NAM David Harold Blackwell Lecture
August 2, 2019 | Cincinnati, Ohio

Omayra Ortega | Emille Davie Lawrence | Edray Herber Goins
Editors

AMERICAN MATHEMATICAL SOCIETY

2010 *Mathematics Subject Classification.* Primary 01-06, 11B83, 97I50, 65D17, 11G05, 05C69, 34G10, 97D40, 05D10, 35J96.

Library of Congress Cataloging-in-Publication Data

Names: Ortega, Omayra, 1978– editor.
Title: The golden anniversary celebration of the National Association of Mathematicians / Omayra Ortega, Emille Davie Lawrence, Edray Herber Goins, editors.
Description: Providence, Rhode Island: American Mathematical Society, [2020] | Series: Contemporary mathematics, 0271-4132; volume 759 | Includes bibliographical references.
LCCN 2020014035 | ISBN 9781470451301 (paperback) | ISBN 9781470461294 (ebook)
Subjects: LCSH: National Association of Mathematicians (U.S.)–Anniversaries, etc. | Mathematics. | AMS: History and biography – Proceedings, conferences, collections, etc. | Number theory – Sequences and sets – Special sequences and polynomials. | Mathematics education – Analysis – Integral calculus. | Numerical analysis – Numerical approximation and computational geometry (primarily algorithms) – Computer aided design (modeling of curves and surfaces – Elliptic curves over global fields. | Combinatorics – Graph theory – Dominating sets, independent sets, cliques. | Ordinary differential equations – Differential equations in abstract spaces – Linear equations. | Mathematics education – Education and instruction in mathematics – Teaching methods and classroom techniques. | Combinatorics – Extremal combinatorics – Ramsey theory. | Partial differential equations – Elliptic equations and systems – Elliptic Monge-Ampére equations.
Classification: LCC QA1 .G65 2020 | DDC 510–dc23
LC record available at https://lccn.loc.gov/2020014035
DOI: https://doi.org/10.1090/conm/759

∞ The paper used in this book is acid-free and falls within the guidelines
established to ensure permanence and durability.
Visit the AMS home page at `https://www.ams.org/`

10 9 8 7 6 5 4 3 2 1 25 24 23 22 21 20

Contents

Foreword by the President of the National Association of Mathematicians vii

Preface ix

The Founding of the National Association of Mathematicians, Inc. (NAM)
JOHNNY L. HOUSTON 1

An overview of mathematical modeling of geometric optics problems involving refraction
HENOK MAWI 21

Women who count: Using the positive narratives of African American women mathematicians to motivate students and build positive mathematics identities
SHELLY M. JONES 39

A constructive proof of Masser's Theorem
ALEXANDER J. BARRIOS 51

On the existence of $C^{(n)}$-almost automorphic mild solutions of certain differential equations in Banach spaces
GASTON M. N'GUÉRÉKATA and GISÈLE MOPHOU 63

A first hitting time approach to finding effective spreaders in a network
FERN Y. HUNT 71

Peer-Led Team Learning and its effect on mathematics self-efficacy and anxiety in a developmental mathematics course
NADIA MONROSE MILLS, ANGELICQUE TUCKER BLACKMON, CAMILLE MCKAYLE, ROBERT STOLZ, and SANDRA ROMANO 93

A discreteness algorithm for 4-punctured sphere groups
CALEB ASHLEY 103

Catalan and Motzkin integral representations
PETER MCCALLA and ASAMOAH NKWANTA 125

The research of seven students at Howard University
NEIL HINDMAN 135

Complex variables, mesh generation, and 3D web graphics: Research and technology behind the visualizations in the NIST Digital Library of Mathematical Functions
BONITA V. SAUNDERS 145

A linear programming method for exponential domination
MICHAEL DAIRYKO and MICHAEL YOUNG 157

The first 50 years of the National Association of Mathematicians, Inc. (NAM)
JOHNNY L. HOUSTON 167

Foreword by the President of the National Association of Mathematicians

"A dream deferred is a dream denied." –Langston Hughes

In 2018, while I was busy at work preparing for NAM's year of 50th Anniversary activities, I received an invitation from the Mathematical Association of America (MAA) to speak at the 2019 Joint Mathematics Meetings. At first I was puzzled, because I did not think I would have anything interesting to say. Then I realized I could give a voice to NAM and to the unsung African American mathematicians who are unknown to the larger mathematical community.

In 1934, Walter Richard Talbot earned his Ph.D. from the University of Pittsburgh; he was the fourth African American to earn a doctorate in mathematics. His dissertation research was in the field of geometric group theory, where he was interested in computing fundamental domains of action by the symmetric group on certain complex vector spaces. Unfortunately, opportunities for African Americans during that time to continue their research were severely limited. "When I entered the college teaching scene, it was 1934," Talbot is quoted as saying. "It was 35 years later before I had a chance to start existing in the national activities of the mathematical bodies." Concerned with the exclusion of African Americans at various national meetings, Talbot helped to found the National Association of Mathematicians (NAM) in 1969. I would tell his story at JMM.

I made the decision to give a talk weaving the history of NAM with the history of African American mathematicians. Indeed, for many people, the dream of becoming a research mathematician was met with many road blocks. Mathematicians such as William Schieffelin Claytor (January 4, 1908 – July 14, 1967), the third African American to earn a doctorate in mathematics, had their careers derailed by towering giants in topology such as Robert Lee Moore. Others, such as Evelyn Boyd Granville (May 1, 1924 –), the second African American woman to earn a doctorate in mathematics, were not allowed to attend banquets at AMS Sectional Meetings due to racist policies by the American Mathematical Society and inaction by the AMS President Saunders MacLane. And others still, such as Vivienne Lucille Malone-Mayes (February 10, 1932 – June 9, 1995), the fifth African American woman to earn a doctorate in mathematics, were refused graduate admission to schools, including Baylor University. Claytor would persist to become the first African American to publish a research paper in mathematics. And Granville would go on to inspire younger African American women to continue in mathematics, such as her undergraduate mentee Etta Zuber Falconer. And Malone-Mayes would return years later to Baylor as their first African American faculty member. The dream to becoming a mathematician was a dream deferred – but not a dream destroyed.

At the 2019 Joint Mathematics Meetings in Baltimore, I gave an address titled "A Dream Deferred: 50 Years of Blacks in Mathematics", on Thursday, January 17. The National Association of Mathematicians began as an organization at the 1969 Joint Mathematics Meetings in New Orleans: On Sunday January 26, seventeen underrepresented minorities met to discuss how they could advocate for inclusion, and no longer let minorities be excluded from the mathematical community. Among those present were a young assistant professor at Baylor University named Vivienne Malone-Mayes, as well as an elder statesman and department chair at Morgan State University named Walter Talbot. Dr. Malone-Mayes had just received her doctorate three years before in 1966. Dr. Talbot could finally see his dream, once deferred, now realized.

NAM hosted many events at the 2019 Joint Mathematics Meetings which showcased African American mathematicians, but the highlight was the NAM Banquet. It took place on Friday January 18 from 6:00 PM - 9:00 PM in Holiday Ballroom 6 on the 2nd Floor of the Baltimore Hilton. Talithia Denese Williams (Harvey Mudd College) gave the annual Cox-Talbot Lecture on "A Seat at the Table: Equity and Social Justice in Mathematics Education". We bestowed NAM's Centenarian Award to Katherine Johnson (NASA); NAM's Lifetime Achievement Award to Mel Currie (Department of Defense); and NAM's inaugural Stephens-Shabazz Teaching Award to Duane Anthony Cooper (Morehouse College). We paid our respects to the late Amassa Courtney Fauntleroy (April 5, 1945 – October 19, 2017). We presented NAM's Founders Award to the five founding members of the organization who are still living: Scott Williams, Harriett Junior Walton, Johnny L. Houston, James A. Donaldson, and Robert S. Smith. And we gave a standing ovation to the recipient of NAM's Golden Anniversary Legacy Award: a 95-year old Evelyn Boyd Granville, who was in the audience!

2019 was an incredible year for the National Association of Mathematicians. We reflected on the history of the organization; we pondered how African American mathematicians persevered over the past 50 years so that we could have as many black mathematicians as we did present at the Joint Mathematics Meetings; we were able to celebrate the lives of those young activists who decided on January 26 that enough was enough; we were able to fellowship with those who are still with us to provide inspiration for the next 50 years; and we were able to welcome the next generation with open arms, give them a platform to showcase their talent, and provide them with welcoming community.

I encourage you to learn more about the history of NAM, the stories of African American mathematicians, and the research done by some very talented underrepresented minorities in the mathematical sciences by reading the articles in this book. I wish to thank NAM's 50th Anniversary Planning Committee for their tireless work in making all of the activities in 2019 possible: Earl Barnes, Farrah Jackson, Duane Cooper, and co-chairs Janis Oldham and Scott Williams. And I wish to thank my co-editors, Emille Davie Lawrence and Omayra Ortega, for taking up the challenge to document the Golden Anniversary of the National Association of Mathematicians. In the words of Marcus Garvey: "Let us with one determination create in our minds today the conditions of the next 50 years."

Edray Herber Goins
President, NAM

Preface

Dear Authors and Esteemed Readers,

It is with a deep satisfaction that I write this preface to the proceedings of The Golden Anniversary Celebration of the National Association of Mathematicians. Celebrations were held throughout the year 2019 at the Joint Mathematics Meetings in Baltimore, Maryland, in the winter, and at the MAA MathFest in Cincinnati, Ohio, over the summer. The National Association of Mathematicians (NAM) is an organization born from the sweat of seventeen founders, whose vision and determination is chronicled in the article, "The Founding of the National Association of Mathematician, Inc." by Dr. Johnny Houston. Their brainchild, NAM, has persisted through several changes in board members through the hard work of many many people.

The papers contained in these proceedings represent the work of a cross section of lecturers, panelists, and keynote speakers who contributed to NAM's Golden Anniversary celebrations. I want to thank my associate editors Emille Lawrence and Edray Goins for their invaluable help and for their patience enduring my innumerable emails. I also want to thank the reviewers: Alejandra Alvarado, Ami Radunskaya, Asamoah Nkwanta, Bonita Saunders, Dawn Lott, Elaine Terry, Erica Walker, Gaston N'Guérékata, Gizem Karaali, Illya Hicks, Jane Gilman, Janis Oldham, John Johnson, Junfei Cao, K.A. Penson, Kasso Okoudjou, Kendra Pleasant, Lakeshia Legette Jones, Matthias Beck, Mel Currie, Nestor Guillen, Olivier Glorieux, Ozlem Ejder, Pamela Harris, Paul Berry, Philip Smith, Robert E. Bozeman, Ron Mickens, Ruth Haas, Ryan Hynd, Shea Burns, Sylvia Bozeman, Torina Lewis, William Goldman, and Yong-Kui Chang.

These proceedings will furnish scientists and historians of the world with a summary of the activities and motivating concepts during the first 50 years of NAM. We thank all authors and participants for their contributions.

Omayra Ortega
Editor-in-Chief
Chair of the Publicity & Publications Committee, NAM

Contemporary Mathematics
Volume **759**, 2020
https://doi.org/10.1090/conm/759/15262

The Founding of the National Association of Mathematicians, Inc. (NAM)

Johnny L. Houston

Abstract. When an organization is established, often there is a great amount of pre-planning and group collaboration before any attempt is made to assemble its perspective founders to formally outline the proper procedures that must be followed to establish the organization. The organization usually follows a prescribed model or similar model that already exists. This was not the case in the establishment of NAM. The right people with the right synergy began to communicate at the right time in the right place and with a similar view of urgency. They said, "let us do it now!" It was spoken; it was agreed; and it was done!

1. Prologue

Sitting at the tutelage of Professor Claude B. Dansby, one of my college mathematics faculty members at Morehouse College, caused me to see some of the wonderful things that one could do with mathematics. Although Prof. Dansby never earned a PhD in mathematics, he inspired nine of his students to earn such a degree, including yours truly. When I began doing graduate study in mathematics, I became curious about which African Americans had earned a PhD in mathematics: Who were they? What were they doing? How challenging was it to be a successful and productive professional mathematician on the national scene? To answer these questions, I attended the Joint National Mathematics Meetings of the American Mathematical Society (AMS) and the Mathematical Association of America (MAA) on January 23-26, 1969 in New Orleans, Louisiana.

The things I observed, learned, and experienced at this Joint Mathematics Meeting in 1969 were daunting. I spent a great deal of time mingling with the few African Americans in attendance and learning their experiences and challenges as African American mathematicians on the national scene and in the Academy. The practice of racism in America had forced most African American mathematicians to be excluded from viable participation at the national scene, and most opportunities had been limited to employment positions in the Academy as faculty members at either Historical Black Colleges and Universities (HBCSs) or Minority Serving Institutions (MSIs).

There were about 20 of us in attendance at the New Orleans Meeting. We decided to meet as a group and have a serious dialogue about the plight and the

2010 *Mathematics Subject Classification.* Primary 01A61, 01A65, 01A70.

current challenges of mathematicians in the American minority populations, especially their lack of visibility and underrepresentation at the national level. Seventeen of us met on Sunday morning January 26, 1969.

2. Dynamics of the times in the larger society

We assembled to assess the situation of underrepresented-minority American mathematicians both nationally and in the Academy. We were all aware, as having observed and experienced, the dynamics of the times in the larger society:

- We were acutely aware of the various exposures and demands in society resulting from the Civil Rights Movement and the killing of Dr. Martin Luther King, Jr. less than a year earlier in April 1968.
- The 1938 Supreme Court's ruling in *Missouri ex rel. Gaines v. Canada (1938)* in which the court ruled that states which provided public higher education for White students also had to provide it for Black students, to be satisfied either by establishing Black colleges and universities or by admitting Black students to previously White-only universities. In 1965 and 1966 the state of Georgia paid money to Clark Atlanta University (then Atlanta University Graduate School) for me, an African American citizen in the state of Georgia, to study for a Master of Science degree in mathematics, and for my not insisting on attending a graduate state research institution in the state of Georgia.
- In 1954, the Supreme Court ruled: in in the landmark case *Brown v. Board of Education of Topeka* (347 U.S. 483) that state laws establishing separate public schools for Black and White students were unconstitutional, and that the practice of segregated schools must be discontinued with all deliberate speed.

These and other Supreme Court rulings were beginning to "open" more graduate study opportunities for underrepresented Americans minorities, especially African Americans. Even though the author had been accepted in a PhD graduate program at Purdue University for the fall of 1969, the author attended the University of Georgia graduate mathematics program in the summer of 1969 in which he was the only person of color in the program. However, in 1969, professional organizations, lucrative professional opportunities in the mathematical sciences in Industry, National Labs, and PhD granting institutions were primarily still deliberately excluding the acceptance and hiring of underrepresented American minorities.

3. Founding principles of the National Association of Mathematicians

At the 1969 meeting in New Orleans, we as a group of USA minority mathematicians decided to begin to identify some proactive leadership actions to bring about more constructive activities and opportunities for underrepresented Americans minorities in the mathematical sciences – especially on the national scene.

Table 2 lists the Initial Seventeen (17) minority American mathematicians which met in New Orleans on Sunday morning on January 26, 1969. They are recognized as the Initial Founders of the National Association of Mathematicians, Inc. (NAM).

FIGURE 1. Some of the early leaders of NAM

Name (Last, First)	January 1969 Institution	1969 Position	1969 Degree	January 2019 Status
Donaldson, James Ashley	University Illinois at Chicago	Faculty	PhD	Deceased
Douglas, Samuel Horace	Grambling College	Faculty	PhD	Deceased
Eldridge, Henry Madison	Fayetteville State College	Faculty	PhD	Deceased
Frazier-Svager, Thyrsa Anne	Central State University	Faculty	PhD	Deceased
Griego, Richard	Univ New Mexico Albuquerque	Faculty	PhD	UNM (Emeritus)
Houston, Johnny Lee	Stillman College/Purdue	Faculty/Grad Stu	MS	Elizabeth City (Emeritus)
Jefferson, Curtis	Cuyahoga Community College	Faculty	MS	Unknown
Malone-Mayes, Vivienne	Baylor University	Faculty	PhD	Deceased
Portis, Theodore	Alabama State University	Faculty	MS	Unknown
Smith, Charles R.	Paine College	Faculty	MS	Unknown
Smith, Robert S.	Penn State University	Grad Student	MS	Miami Univ (Emeritus)
Stubblefield, Beauregard	Texas Southern University	Faculty	PhD	Deceased
Thaggert, Henry.	Jarvis Christian College	Faculty	MS	Unknown
Talbot, Walter Richard	Morgan State College	Faculty	PhD	Deceased
Valez-Rodriquez, Argelia	Bishop College	Faculty	PhD	GVT (Retired)
Walton, Harriet Rose Junior	Morehouse College	Faculty	MS	Morehouse (Emeritus)
Williams Scott W.	Lehigh University	Grad Student	MS	SUNY Buffalo (Emeritus)

FIGURE 2. Names and Current Status of the 17 Founders of NAM

This group of 17 persons, all underrepresented American minorities in the mathematical sciences, are given credit for bringing about the existence of NAM as a professional organization. When the group met they discussed many issues and posed many questions, such as:

Where do we go from here?
If not now to begin change, then when?
If not us, then who?
What should be the primary visionary goal of these changes?

Our primary vision was that underrepresented-minority American mathematicians would themselves serve as a positive force to make a difference of their status in the mathematical sciences community. Our goal was to embrace the Kwanzaa principle of *Kujichagulia* – Swahili for "self-determination". We decided not to allow others to define us. Indeed, it was our intention to promote

- excellence in the mathematical sciences,
- the development of underrepresented American minorities in the mathematical sciences,
- "openness" in the participation of the scholarly arena, and
- inclusion and not accept exclusion.

It was our intention to define ourselves holistically as we participated around the world at conferences, sat at meeting tables, and participated in scholarly arenas in the mathematical sciences community. In less than one year, these 17 persons had been joined by scores of other like-minded mathematicians. And in less than two years, this Initial Force of 17 persons had become the National Association of Mathematicians, Inc. (NAM).

4. NAM's early years

Walter R. Talbot, one of the 17 founding members of NAM, was more senior than most as he was one of the first African Americans to earn a PhD in mathematics. He secured funds for many of the early meetings of NAM. These meetings continued the discussions that were initiated at the New Orleans meeting in 1969. In his own generous and professional way, Professor Talbot never sought, requested, or desired special recognition for his many contributions. Neither did he seek a leadership position in the young organization. Professor Talbot was content to be a positive catalyst for early growth: he was content to be a driving force in the establishment of NAM.

There were many others played who played important roles in the early years of the organization. Frank James was the first elected president of NAM. Benjamin J. Martin at Morehouse College and Etta Falconer at Spelman College both received their doctoral degrees in mathematics in 1969; they both did a lot of work behind the scenes. Even so, it was Professor Talbot who was the "wind beneath NAM's wings" during NAM's formative years.

5. Influence on other organizations

Many people believe that NAM is focused on African American mathematicians, but NAM began in the early 1970s with requests that all underrepresented American mathematicians of color join together as a collective voice. NAM wanted all to become more aware of the plight of underrepresented American minority mathematicians in the larger mathematical sciences community at the national scene. Indeed, NAM was founded in the interest of inclusion, openness, fairness, progress, and improvement within in the mathematical sciences community in the USA. NAM wanted the issues of awareness and lack of recognition to be both seen and known by all.

NAM advanced the issues of awareness of racism in the mathematical sciences for underrepresented American minorities at the national scene. The major mathematical sciences organizations, such as AMS and MAA, began to examine their policies and practices. In 1971, two years after the beginning of NAM, women mathematicians in the United States started a new organization: the Association of Women in Mathematics, Inc. (AWM). In 1973, the Latinx/Hispanic and Native American communities started a new organization: the Society for Advancement of Chicanos/Hispanics and Native Americans in Science (SACNAS). NAM even raised the issue of awareness at such a high level that governmental agencies and private industries alike began to discuss support for the establishment of an: Office of Awareness for the Mathematical Sciences in Washington, DC.

More information on these formative years for NAM can be found in [**1**] and [**3**].

6. 1969: A year of challenge and hope

In 1969 the larger mathematical sciences community viewed the Joint Mathematics Meeting in New Orleans as a great success. According to [**6**], there were 4,811 persons registered for the 1969 Joint Mathematics Meetings, which included 3,084 members of the American Mathematical Society; it was one of the largest gathering up to that point. The larger mathematical sciences community appeared to be happy with how things were.

However, the representation of under-represented American minorities in the larger mathematical sciences community appeared to be of no concern and non-existent, for all practical purposes in 1969. There appeared to be of no concern that the USA was a diverse country with at least 10% of the population being African Americans. At the time, less than 1% of the mathematical scientists were African Americans. The vast majority of the mathematics departments in research universities had never admitted or awarded a single doctoral degree to an African American. It appeared that many of these departments had no plan to do so – some even stating explicitly that they never would.

Up through 1969, there were other challenges even creating a pipeline to generate more African Americans with doctorate degrees in the mathematical sciences: There appeared to be no concern that 99% of research universities had no African Americans on their faculties in mathematics. And there were several major deaths of African American mathematicians: Elbert Frank Cox, the first African American to receive a PhD in mathematics, and Joseph Pierce, the sixth African American to earn a PhD in mathematics, died in 1969. Dudley Woodard, the second African American to receive a PhD in mathematics, died in 1965. William Claytor, the third African American to receive a PhD in mathematics, died in 1967. All four passed away before there could be a critical mass of 100 African Americans who would received a PhD degree in mathematics.

I was a graduate student at Purdue University pursuing my own doctorate degree in mathematics in 1969, and I hoped to eventually be counted within this critical mass. But I had some serious questions: What would it take to change the conditions African Americans faced in the mathematical sciences community? Where could one find hope in light of these circumstances? I would eventually realize that there was hope from implications of decisions made by the Federal Courts of the USA; hope in the wave of changes being brought about nationally by the Civil Rights Movement; hope in the evolution of NAM

There was hope in that David Blackwell, an African American mathematician, had reached his 50th birthday in 1969 – and was a trail-blazing tenured full professor in statistics at UC Berkeley who had been elected as a Fellow to the National Academy of Science. (See my article [**4**].)

There was also hope in 1969, in that my 50 year-old mother with a 4th grade education who was head of household in a one parent family with four children, had produced the family's first college graduate who had been accepted to study in a PhD program in the mathematics department in a research university. And there was hope in that I decided to marry my high school sweetheart in 1969 and carry her with me to the mid-west, as I pursued a doctoral degree in mathematics. See, when I arrived at Purdue in 1969, Black students were demanding the establishment of Black Cultural Center (BCC). Students were insisting on changes. And that gave me even more hope.

The winds of positive change were beginning to blow stronger in the direction of hope than in the direction of dismay for African Americans who wanted to pursue careers in mathematics.

7. Profiles of NAM's founding members

NAM celebrated its 50th Anniversary at the 2019 Joint Mathematics Meetings. We held a series of events in Baltimore, Maryland from January 16-19, 2019. We honored its 17 founders during the annual NAM Banquet on Friday, January 18 from 6:30 PM - 8:30 PM. Only five (5) of the original 17 attended this celebration: James Donaldson, Johnny L. Houston, Robert Smith, Harriet J. Walton, and Scott W. Williams. NAM was able to provide profiles for all but 4 founders for the audience who attended NAM's Banquet in January 2019. We reproduce those profiles in the remaining pages of this article. Unfortunately we have not been able to locate any information for the following founders:

- Curtis Jefferson (Cuyahoga Community College)
- Theodore Portis (Alabama State University)
- Charles R. Smith (Paine College)
- Henry Thaggert (Jarvis Christian College)

NAM would like to salute all the founders for laying such an excellent foundation for the professional edifice we call NAM!

7.1. James Ashley Donaldson (1941 - 2019). James Ashley Donaldson (Figure 3a) was born in 1941 in Madison County, Florida. He earned a BA in mathematics from Lincoln University in Pennsylvania in 1961; and a MS from the University of Illinois at Urbana-Champaign in 1963. Donaldson received his PhD in mathematics from the University of Illinois at Urbana-Champaign in 1965, writing a doctoral dissertation entitled *Integral Representations of the Extended Airy Integral Type for the Modified Bessel Function* under the supervision of Ray

(A) James Ashley Donaldson (1941 - 2019)
Emeritus, Howard University

(B) Samuel Horace Douglas (1924 - 1989)
Deceased, Grambling State College

(C) Henry Madison Eldridge (1924 - 2010)
Deceased, Fayetteville State College

(D) Thyrsa Frazier-Svager (1930 - 1999)
Deceased, Central State University

FIGURE 3. 4 of the 17 Founding Members of NAM

G. Langebartel. Donaldson's scholarly interests in differential equations continued well into his retirement, and his research publications include numerous papers in analysis, differential equations, and applied mathematics. He lectured widely on his research in North America, Africa, Asia, and Europe. Donaldson was also interested in the History of Mathematics, Mathematics Education and the training of Mathematics Teachers.

Donaldson entered Lincoln University near Oxford, Pennsylvania in 1957. The energetic and clear mathematics teaching of Professor James Frankowsky was a great contributing factor in Donaldson's decision to major in Mathematics. This same person encouraged Donaldson to pursue a graduate degree in mathematics, and Donaldson earned two. After graduating, Donaldson held appointments at Southern University during the summers of 1964, 1965, and 1966; Howard University from 1965-1966, the University of Illinois at Chicago from 1966-1971, and the University of New Mexico from 1969-1971. He held visiting positions at the Courant Institute of Mathematical Science; the University of Victoria in British Columbia, Canada; the University of Ferrara in Italy; and Duke University.

Donaldson returned to Howard University in 1971. During his tenure as department chair from 1972-1990, the Howard University Mathematics Department underwent a transformation ushering in a strong research program that justified the development and inauguration of the first doctoral degree program in mathematics at an Historically Black College or University. This program has become a major producer in America of African American holders of a PhD in Mathematics.

Donaldson has held national offices in several professional organizations, including serving as member of the Council of the American Mathematical Society (AMS), the second Vice-President of the Mathematical Association of America (MAA), and editor of the NAM Newsletter. (See [**6**].) Donaldson has also served as a consultant to the National Science Foundation (NSF), the National Research Council (NRC), and the Educational Testing Service (ETS). Donaldson was a former President of the Geneneral Alumni Association of Lincoln University; member of the Board of Trustees at Lincoln, and was appointed Acting President at Lincoln during the fall of 1998. He also received an Honorary Doctorate from Lincoln in 2019.

7.2. Samuel Horace Douglas (1924 - 1989). Samuel Horace Douglas (Figure 3b) was born in 1924. He received a B.S. in mathematics from Bishop College in 1948, and his MS in mathematics in from Oklahoma State University in 1959. Douglas continued to pursue graduate studies while teaching at the collegiate level. In 1967, he was awarded the PhD in mathematics by Oklahoma State University with a doctoral dissertation entitled *Convexity Lattices related to Topological Lattices and incidence Geometries.*

Douglas, a founder of NAM, spent many years of his professional career as Chair of the Department of Mathematics at Grambling State College (now Grambling State University) in Louisiana. Douglas was very active as member of NAM and from 1977 - 1983: he even served as NAM's president. One of his more notable contributions was that of helping to plan, organize, and implement NAM's Tenth Anniversary Celebration, from March 30-31, 1979 in Boulder, Colorado. This workshop-conference was supported by NAM's first major grant as awarded by the Environmental Research Labs of the National Oceanic and Atmospheric Administration (NOAA). There were some over 100 NAM members representing some 35 Historical Black Colleges and Universities (HBCUs) at this meeting. (NAM member and founder Beauregard Stubblefield, who worked for NOAA at the time, helped NAM to secure the grant.)

7.3. Henry Madison Eldridge (1924 - 2010). Henry Madison Eldridge (Figure 3c) was born in Montgomery, Alabama in December 1924. He earned a B.S. from Alabama State College, an M.A. from Columbia University, and a PhD from the University of Pittsburgh in 1956. He was Professor of Mathematics at North Carolina State University at Fort Bragg from 1966-72. In 1972, Eldridge went to Fayetteville State University, and remained there until retirement. For a while, Eldridge served as the Associate Vice-Chancellor for Academic Affairs. The Mathematics Department at Fayetteville State University is named in his honor.

7.4. Thyrsa Anne Frazier-Svager (1930 - 1999). Thyrsa Anne Frazier-Svager (Figure 3d) was born Thyrsa Anne Frazier in June 1930 in Wilberforce, Ohio. Frazier graduated from Antioch College in 1950 with a major in mathematics and a minor in chemistry. She placed in the 99th percentile in the Princeton Senior Student Examination. Frazier was one of only four black students at Antioch: one of the others was Coretta Scott King, with whom she was friends. She gained a bachelor of arts degree from Antioch in 1951, going on to gain a master's in 1952 and PhD from the Ohio State University in Columbus in 1965. At Ohio State, she wrote a doctoral dissertation entitled *On the product of absolutely continuous transformations of measure spaces* under the supervision of Paul Reichelderfer. Frazier was the seventh African-American woman to earn a PhD in mathematics.

Frazier-Svager worked for a year at Wright Patterson Air Force Base in Dayton, before teaching at Texas Southern University in Houston. In 1954, she joined the faculty of Central State University in Wilberforce. In 1967, Frazier-Svager was appointed Chairman of the department of mathematics. She was awarded tenure in 1970. She was Provost and Vice President for Academic Affairs at Central State University when she retired in 1993. Frazier-Svager spent the summer of 1966 in Washington, DC as a systems analyst at NASA; she was then a visiting faculty at the Massachusetts Institute of Technology (MIT) in both 1969 and 1985. In March 1995, she returned for a short time to Central State University as Interim President. While on the Central State faculty, Frazier met Aleksandar Svager, a Holocaust survivor from Yugoslavia and physics professor at Central State. They married in June 1968 at her parents' home. Thyrsa Frazier-Svager died on July 23, 1999. Frazier-Svager was honored with an Honorary Doctor of Humane Letters by Central on her retirement, and she was inducted into the Hall of Fame in Greene County, Ohio.

Frazier-Svager wrote two books, Central State University's *Modern Elementary Algebra Workbook* (1969) and *Essential Mathematics for College Freshmen* (1976). Frazier-Svager was also a member of Beta Kappa Chi, the National Association of Mathematicians (NAM), and the Mathematical Association of America (MAA), and participated in the meeting that founded the National Association of Mathematics in 1969.

7.5. Richard Griego (1961 -). Richard Griego is a native New Mexican. He earned his B.S. in mathematics at the University of New Mexico in 1961, and his PhD in mathematics at the University of Illinois at Urbana-Champaign in 1965. His doctoral dissertation, entitled *Local Times for Markov Processes*, was supervised by Lester L. Helms.

Griego taught at the University of California at Berkeley for one year, at the University of New Mexico for 26 years, and at Northern Arizona University for five years. He served as chair of the mathematics departments at both the University of New Mexico and at Northern Arizona University. He was the director of several programs in mathematics and STEM education – programs which were especially geared for low-income and minority students. After his retirement in 1997, Griego published articles on Chicano language and identity, on Chicano participation in the sciences, and on colonialism.

7.6. Johnny Lee Houston (1941 -). Johnny Lee Houston (Figure 4a) was born on November 19, 1941 in Sandersville, Georgia; and was the youngest son of the late Mrs. Catherine Houston Vinson.[1] Houston's Current Research interests are Computational Science, Combinatorics, Discrete Geometry and the History of Mathematics. He wrote a book, *The History of NAM: The first 30 years*, in 2000.

[1]The author is celebrating the Centennial Year of his mother, Mrs. Catherine Houston Vinson, in 2019.

(A) Johnny Lee Houston (1941 -)
Emeritus, Elizabeth City State University

(B) Vivienne Malone-Mayes (1932 -1995)
Deceased, Baylor University

(C) Robert S. Smith (1941 -)
Emeritus, Miami University of Ohio

(D) Beauregard Stubblefield (1923 - 2013)
Deceased, Texas Southern University

FIGURE 4. 4 of the 17 Founding Members of NAM (continued)

He plans to publish another book by 2020 entitled *Profiles of African American Mathematicians.*

Houston earned three degrees in mathematics: a BA from Morehouse College in 1964; a MS from Clark Atlanta University in 1966 (at the time it was known as Atlanta University); and PhD from Purdue University in 1974. His doctoral dissertation was entitled *On the Theory of Fitting Classes in Certain Locally Finite Groups*, and was supervised by Purdue professor Eugene Schenkman. Houston was the first Manager/Director of Purdue's Black Cultural Center from 1971-72.[2] Houston did additional study at L'Universite Strasbourg in France from 1966-1967, as well as at the University of Georgia during the summer of 1969.

Houston is Professor Emeritus at Elizabeth City State University (ECSU) in Elizabeth City, North Carolina. During his tenure from 1984-2010, Houston has served as Vice Chancellor for Academic Affairs; Senior Research Professor; Director of a Computational Science - Scientific Visualization Center; and Director of a Global Leadership Academy. Houston has received ECSU Foundation Award as well as ECSU Chancellor Award, and the annual Johnny L. Houston Mathematical Sciences Colloquium (which takes place every October at Elizabeth City State University) was established in his honor.

He served as a Visiting Scientist at various national labs: NASA Langley Research Center, Lawrence Livermore National Laboratory (LLNL), Argonne National Laboratory, Oak Ridge National Laboratory, and the National Center for Atmospheric Research (NCAR). Houston is a Life Member of the National Association of Mathematicians (NAM), the Society for Industrial and Applied Mathematics (SIAM), the Mathematical Association of America (MAA) – where he served on the Board of Governors from 1992-1995, the Purdue National Alumni Association, the Clark Atlanta National Alumni Association, the Morehouse College National Alumni Association, and the National Association for the Advancement of Colored People (NAACP). Houston has been a member of the American Mathematical Society (AMS), the Association for Computing Machinery (ACM), the Institute of Electrical and Electronics Engineers (IEEE), the Association of Computer/Information Sciences and Engineering Departments at Minority Institutions (ADMI) – serving as the second Vice-President from 1990-1994, the Human Resources Advisory Committee at the Mathematical Sciences Research Institute (MSRI's HRAC) from 1993-1998, the Advisory Committee for the North Carolina Super Computer Center from 1994-2003, the Benjamin Banneker Association (BBA), and an Advisory Committee for the Purdue University Mathematics Department from 1999-2002.

Houston received the QEM Award of Excellence in Teaching Mathematics and Science, University of North Carolina Board of Governor's Teaching Excellence Award, and the North Carolina Governor's Award for Outstanding Volunteer Services. He was selected to establish the Professor Claude B. Dansby Display at Morehouse College, and received the the Purdue University Black Cultural Center (BCC) Pioneer Award. Houston was also guest at the White House for a State Dinner in September 2008.

Houston was nationally selected as a Science History Maker; his interview was recorded for the United States Library of Congress. He has received some $15

[2]The Black Cultural Center (BCC) at Purdue University is celebrating its 50th Anniversary in 2019.

million in grants for various scholarly activities. He has produced some 40 publications, including seven books (some in French) of which he is author, co-author or editor-in-chief. He has made scores of scholarly presentations as an Invited Speaker while traveling globally: he has visited 6 continents, 70 different countries (25 of which are in Africa), and all 50 states in the USA. He has taught and guided many students in scholarly pursuits at the B.S., MS and PhD levels while teaching at various institutions such as Clark Atlanta University (Atlanta University), Elizabeth City State University (ECSU), Fort Valley State University, Morehouse College, Purdue University, Savannah State University, and Stillman College.

Houston is also Executive Secretary Emeritus of the National Association of Mathematicians, serving as the first active Executive Secretary from 1975-2000. He is also a founder of NAM, being one of the attendees of the initial meeting of NAM on January 26, 1969. Houston has received NAM's Lifetime Achievement Award as well as NAM's Founder Award.

7.7. Vivienne Lucille Malone-Mayes (1932 - 1995). Vivienne Lucille Malone-Mayes (Figure 4b) was born in February 1932 in Waco, Texas. She was the 13th African American woman to earn a PhD in mathematics. She received her B.S. from Fisk University in 1952, and her MS from Fisk University in 1954. While at Fisk, Malone-Mayes was taught by Evelyn Boyd Granville, the second African American Woman to earn a PhD in mathematics.

Malone-Mayes then returned to Waco and served as chair of the mathematics department at Paul Quinn College from 1954-1961. She attempted to enroll at Baylor University in Waco as a doctoral student, but was denied admission because of her race. Malone-Mayes was admitted instead to the to the University of Texas at Austin, where she earned a PhD in mathematics in 1966. She was just the second African American to earn a doctorate degree at that University. Her doctoral dissertation, entitled *A Structure Problem in Asymptotic Analysis*, was supervised by Don Edmonson. She published her dissertation in the Proceedings of the American Mathematical Society in 1969.

Malone-Mayes returned to Waco in 1966 where, ironically, she became the first African American faculty member at Baylor University, the institution that rejected her graduate student application just five years earlier. Malone-Mayes spent the rest of her career at Baylor University, until she became ill and retired in 1994. In her later years, Malone-Mayes's research interests moved towards summability theory. In 1980, she published a joint paper entitled *Some properties of the Leininger generalized Hausdorff matrix* with B. E. Rhoades; it appeared in the Houston Journal of Mathematics.

Malone-Mayes served on the Board of Directors of the National Association of Mathematicians (NAM), and was the first black woman elected to the Executive Committee of the Association of Women in Mathematics (AWM). She was also director of the Youth Choir and served as the organist at New Hope Baptist Church from 1960-1975. Malone-Mayes died on June 9, 1995 in Waco, Texas. As the 13th African American woman to receive a PhD in mathematics, Malone-Mayes made an enormous impact on numerous organizations dedicated to addressing problems impacting women and people of color.

7.8. Robert S. Smith (1941 -). Robert S. Smith (Figure 4c) was born in 1941 in Baltimore, Maryland. He earned his B.S. from Morgan State College (now

Morgan State University) in 1963. He then attended Pennslvania State University where he was able to earn his MS in 1967 and his PhD in 1969. His doctoral dissertation was entitled *On the Distributivity of Filter and Ideal Lattices.*

Since 1969, Smith has been on the faculty of Miami University of Ohio in the department of Mathematics and Statistics. However, his career has been punctuated with visiting positions at the University of North Carolina at Chapel Hill, Butler University of Indianapolis, and LaTrobe University in Melbourne, Australia.

Smith has been involved in a number of innovative computer-based teaching projects. These include teaching undergraduate mathematics with a spreadsheet, and teaching abstract algebra using cooperative learning (i.e., group work) and ISETL ("Interactive SET Language", a mathematical programming language). Smith is an expert in the area of teaching mathematics, and is a frequent presenter at regional, national, and international conferences on technology in collegiate mathematics. He has been nominated for over ten outstanding teacher or distinguished educator awards for his dedication to his students and the pedagogical effort. In 1994, he was awarded the Mathematical Association of America Ohio Section Award for Distinguished College or University Teaching of Mathematics.

Smith is also an international educational consultant. From 1995 to 1998, he was an Assistant Examiner and the External Advisor in Higher Level Mathematics for the International Baccalaureate Organization in Cardiff, Wales. Smith served as the advisor to the Ohio Delta Chapter of Pi Mu Epsilon, the national mathematics honor society, for six years. During his tenure as advisor, the Ohio Delta Chapter of Pi Mu Epsilon became one of the largest and most active chapters in the nation. In 1992, Smith organized and held the first Joint Pi Mu Epsilon/Mathematical Association of America (MAA) Student Chapters Meeting at Miami University. He has been a national officer in Pi Mu Epsilon since and served as President. Smith has served on various regional and national committees in the Mathematical Association of America (MAA).

7.9. Beauregard Stubblefield (1923 - 2013). Beauregard Stubblefield (Figure 4d) was born on July 31, 1923 in Navasota, Texas. As a child, Stubblefield gained a desire to pursue mathematics from his father. In 1943, Stubblefield earned a B.S. in 1943 and a MS in 1945 from Prairie View College (now known as Prairie View University) in Texas; and a PhD in 1980 from the University of Michigan.

From 1961-1967, Stubblefield taught at Oakland University in Michigan. From 1969-1971, he was the Director of Mathematics in the Thirteen College Curriculum Program. From 1976-1981, he served as Mathematician/EEO Manager at the Department of Commerce in Boulder, Colorado. In 1992, Stubblefield retired from the U.S. Department of Commerce with GERL/ERL/NOAA. Stubblefield lived and traveled to many places during his lifetime, but his life ambition was to work in number theory.

Stubblefield was there to establish NAM in 1969. In 1978, he secured NAM's first major grant from NOAA. In 1980, he was the lead research mathematician who was an author of the volume *Black Mathematicians and Their Works.* In 1994, Stubblefield received NAM's Distinguished Service Award; and, in 2000, Stubblefield became the eighth recipient of NAM's highest award, the Lifetime Achievement Award.

(A) Walter Richard Talbot (1909 - 1977)
Deceased, Morgan State University

(B) Argelia Velez-Rodriguez (1936 -)
Retired

(C) Harriet Rose Junior Walton (1933 -)
Emeritus, Morehouse College

(D) Scott W. Williams (1943 -)
Emeritus, SUNY Buffalo

FIGURE 5. 13 of the 17 Founding Members of NAM (continued)

7.10. Walter Richard Talbot (1909 - 1977). Walter Richard Talbot (Figure 5a) was born in Pittsburgh, Pennsylvania on December 9, 1909. He attended the University of Pittsburgh and received his A.B., M.A., and PhD degrees in mathematics from that institution in 1931, 1933 and 1934, respectively. In 1934, he accepted an Assistant Professorship in the mathematics Department at Lincoln University in Missouri. He remained there until 1963, moving through the ranks to professor during his tenure. While at Lincoln, he held several administrative positions, including Chairman of the mathematics department from 1940 - 1963; Dean of Men from 1939 - 1944; Registrar from 1946-1948; and Acting Dean of Instruction from 1955-1957. In 1963, Talbot moved to Morgan State University (formerly Morgan State College) as Chair and Professor of Mathematics. He retired from there in 1977.

Talbot's scientific interests were in mathematical and numerical analysis and computer science. His doctoral dissertation was entitled *Fundamental Regions of S_6 for the Simple Quaternary G_{60}, Type I.* During his career, Talbot was concerned about the teaching of mathematics and computer science. He served the Mathematical Association of America (MAA) in several capacities, and played a major role in the founding of NAM. In 1978, NAM honored him "in memoria" at a luncheon at Morgan State University, and the HBCU has named a scholarship in his honor. In 1980, NAM established the Cox-Talbot Address which is given annually at NAM's national meeting. His contributions to teaching and his service to the mathematical sciences community were very pronounced.

Perhaps more than any other one individual, it was Walter R. Talbot whose activities, guidance and leadership lead to the establishment of NAM. He took the leadership to seek funds, organize meetings and serve as facilitator to bring persons together to discuss issues during the critical years when NAM was being formally established, beginning on January 26, 1969 at the Joint Winter Mathematics Meetings in New Orleans. In October 1969, Talbot organized and conducted the Morgan State Conference. Although the purpose of this conference was designed to concern itself with basic mathematics curricula for TBI's (Traditionally Black Institutions – an older term for Historically Black Colleges and Universities [UBCU]), much of the conference was a follow-up of the concerns that had been voiced in New Orleans. Talbot chaired another related meeting in Memphis, Tennessee in April 1970. In August of that same year, Talbot co-directed a meeting in Laramie, Wyoming called the Conference on Mathematics at Developing Colleges (CMDC). This conference was an effort of the Committee on Assistance to Developing Colleges (CADC), and served as a follow-up to the Morgan State Conference in 1969. At CMDC, the concept of formally organizing an organization for underrepresented-minority American mathematicians was enthusiastically supported by those present. The body selected as the name of the organization: the National Association of Mathematicians (NAM). The body agreed upon a slate of *pro tempore* officers, and recommended memberships for individuals and institutions alike.

For his status as a scholar and his many efforts that lead to the founding and establishment of NAM we are grateful that Walter R. Talbot was there to guide us.

7.11. Argelia Velez-Rodriguez (1936 -). Argelia Velez (Figure 5b) was born in Cuba in 1936. Her father was Pedro Velez; he worked in the Cuban Congress under Fulgencio Batista. Velez's family was Roman Catholic. She attended Roman Catholic schools, and showed her ability in mathematics at an early age: she won an arithmetic competition at her school when she was nine years old. Velez married Raul Rodriguez in 1954.

Velez-Rodriguez earned a B.S. in mathematics at the Marianao Institute in 1955. She continued on to graduate studies at the University of Havana. Most of her professors were women with doctorates. Only around 10% of the Cuban population was Black. In 1960, she became the first Black woman to receive a doctorate of science in mathematics in Cuba. Her doctoral dissertation was entitled *Determination of Orbits Using Talcott's Method.*

Velez-Rodriguez emigrated from Cuba to the United States in 1962 where she began teaching mathematics in Texas. Her husband, however, was unable to leave Cuba with her. He was forced to remain in Cuba until he was able to join her three years later. Velez-Rodriguez joined Bishop College in Dallas in the 1970's. In 1972, Velez-Rodriguez became an American citizen. At Bishop College she was a professor, and she served as chair of the Department of Mathematical Science from 1975-1978.

It was in 1970 that she first became involved with the science education programs of the National Science Foundation (NSF), a move which marked the start of what would eventually become her life's work. In 1979, Velez-Rodriguez took leave of absence from Bishop College to become a program manager with the Minority Institutions Science Improvement Program (NSF) in Washington, DC. She did not return to Bishop; however, in 1980, she became director of the Minority Science Improvement Program at the U.S. Department of Education, from which she retired.

7.12. Harriet Rose Junior Walton (1933 -). Harriet Rose Junior Walton (Figure 5c) was born in Claxton, Georgia in September 1933. She earned her a B.S. from Clark College (now Clark Atlanta University) in mathematics under the tutelage of Joseph J. Dennis. She earned her MS at Howard University in 1954, where she served as a Teaching Assistant and a Research Assistant under David Blackwell; Blackwell subsequently directed Walton's MS thesis. Walton also studied under Elbert Cox, George Butcher, and William W. S. Claytor.

Upon graduation from Clark, Walton taught at Hampton Institute (now Hampton University) from 1954-1955. She was offered a graduate fellowship as a teaching assistant at Syracuse University, where she enrolled from 1955-1957. Her mentor, Dr. Abe Gelbart, encouraged her to stay two years and complete a master's degree rather than leave after one year in order to get married. Walton graduated from Syracuse with a M.A. in mathematics in 1957. Harriett returned to Hampton Institute as an Assistant Professor from 1957-1958. In June 1958, she married James Walton. Walton moved to joined the faculty of Morehouse College in September 1958 during the presidency of Dr. Benjamin Elijah Mays, Walton's marriage led to four children and several grandchildren.

As a college professor, Walton was primarily interested in getting the best possible graduate education for herself, and in giving the best possible college education to her students. Walton took classes at the Georgia Institute of Technology (Georgia Tech) from 1964-1966, and participated in summer courses at Emory University in 1966. Walton earned more degrees later in life: she earned a PhD in mathematics education from Georgia State University in 1979, and she earned a MS in computer science from Atlanta University (now Clark Atlanta University) in 1989.

Throughout her career, Walton served as a part-time lecturer at other institutions in the Atlanta University Center, at Georgia State University, and at Atlanta Junior College. Walton was president of the National Council for Teachers of Mathematics (NCTM), a speaker at various annual and regional meetings of NCTM; and served as either a committee member or officer of various professional organizations such as the American Mathematical Society (AMS), the Mathematical Association of America (MAA), the National Association of Mathematicians (NAM), and the National Science Foundation (NSF). Walton is also a published author. Among her former students are mathematicians Geraldine Darden, Benjamin Martin, and Johnny Houston.

Walton has received many accolades over her career. She was a Fulbright Fellow, and travelled to both Ghana and Cameroon in West Africa. Walton was awarded Teacher of the year at Morehouse in 1990. She was named a United Negro College Fund Fellow at Georgia Tech from 1964-1965; an NSF Faculty Fellow at Georgia Tech/Emory University from 1965-1966; and a U.N.C.F./Dana Fellow at Georgia State University from 1975-1977. Walton is listed in several Who's Who as well as other publications: she was listed in "Who's Who Among Students in American Colleges and Universities" in 1951; "Outstanding Educators of America" in 1971; ''Who's Who of American Women" in 1974; "Personalities of the South" in 1974; "Men and Women of Science" in 1982; "Who's Who in Georgia" in 1982; and"Who's Who Among Black Americans" in 1985. Walton is a Golden Year Member of Delta Sigma Theta Sorority, Inc. She serves as a Deacon at Providence Baptist Church. In 1984, Walton was elected to the Phi Beta Kappa, Delta of Georgia Chapter.

In May 2000, Walton retired from Morehouse College after 42 years of service. Walton's legacy is still impactful at Morehouse: since 2001, the Harriett J. Walton Symposium on Undergraduate Mathematics Research occurs each spring at the college, and the H. J. Walton Senior Prize is given each year at Morehouse.

Walton is a Lifetime Member of NAM, and she has received two Distinguished Service Awards from the organization. For nearly 10 years, Walton served as Secretary-Treasurer of NAM.

7.13. Scott W. Williams (1943 -). Born in New York City in 1943 and raised in Baltimore, Scott Williams (Figure 5d) comes from a line of academics and political activists. His interest in the mathematical sciences was discovered in grade school, where there were multiple grades in one room – so he could learn what the higher grades learned after doing his own work. This interest was further cultivated while in high school when a research mathematician was willing to work with Williams on advanced mathematics.

At the age of 12, Williams's mother took him to see the Massachusetts Institute of Technology (MIT) during a family trip to Boston. After his mother's description of the Institute as a "great place of mathematical learning", Williams replied, "Mom, I will get a PhD here in mathematics." Although Williams scored extremely well on what is now the SAT, he did not receive a scholarship to M.I.T. Instead, Williams attended Morgan State College (now Morgan State University), along with other future African American mathematics professors Earl Barnes and Arthur Grainger. Barnes, Grainger, and Williams became involved students in Dr. Clarence Stephen's mathematics learning program, now known as the "Morgan-Potsdam Model". By the time Williams received a B.S. in mathematics from Morgan State in 1964, he had solved four advanced problems in the Mathematical Monthly; and had co-authored two papers on non-associative algebra with his undergraduate advisor, Dr. Bohun Volodymir-Chudyniv.

Williams went on to receive his MS and PhD in Mathematics from Lehigh University. After a postdoctoral position at Pennsylvania State University, Williams joined the State University of New York at Buffalo (SUNY Buffalo). Williams has also spent some time as a Fulbright Professor at Charles University in the Czech Republic.

While Williams has many accolades, his most proud academic accomplishments include invitations to speak in foreign countries about his research such as New Zealand, China, England and Poland; winning the New York Chancellor Award for Excellence in Teaching in 1982; and being named in 2004 one of the fifty most important Blacks in Research Science. In contrast, his most proud personal accomplishments are his three wonderful daughters, each of whom has given him wonderful grandsons; and, meeting two women, his first and second wives, who did not crush his mathematical interests with demands. To this day, he continues to participate in the Topology and Set Theory mathematical communities by willingly driving two hours each way weekly in order to participate in research seminars.

Williams is perhaps best known for creating and maintaining the website Mathematicians of the African Diaspora (the "MAD Pages"); see [**5**]. Not only was Williams a founder of NAM, but he was also one of NAM's most impactful editors of its Newsletter for more than a decade. (See [**6**].)

8. Epilogue

It was both ironic and befitting that the first African American to earn a PhD in mathematics, Elbert Frank Cox, died in 1969, the same year that NAM was founded. Cox ignited a flame that the founders of National Association of Mathematicians declared would never be extinguished. The flame would continue to "Grow and Glow" for all of humanity to know that excellence and acceptance in mathematics would continue to be available to all, especially for persons in the population of underrepresented American minorities.

In recognition of Elbert F. Cox's pioneering efforts and Walter R. Talbot's diligence and superior leadership in the early days of NAM, the organization honored these two men in 1980 when NAM established the Cox-Talbot Lecture. This invited address is held annually at NAM's National Meeting during the Joint Mathematics Meetings in January of each year.

Photo Credits

Figures 1, 3A, 3B, 4A, 4B, 4D, and 5C are from the personal library of Dr. Johnny L. Houston. Figures 3C, 4C, 5A, and 5B are courtesy of Virginia Newell, Joella Bigson, Waldo Rich, and Beauregard Stubblefield, Mathematicians and their Works (1980). Figure 3D is from Wikimedia Commons. Figure 5D is courtesy of Scott Williams (personal photo).

References

[1] Johnny L. Houston. *The History of the National Association of Mathematicians (NAM), The First 30 years.* NAM (2000).

[2] Johnny L. Houston. Private Correspondences and Photographic Collection.

[3] Johnny L. Houston, *Ten African American pioneers and mathematicians who inspired me*, Notices Amer. Math. Soc. **65** (2018), no. 2, 139–143, DOI 10.1090/noti1639. MR3751310

[4] Johnny L. Houston, *A centennial year (2019) reflection on the life and contributions of mathematician David H. Blackwell (1919–2010)*, Notices Amer. Math. Soc. **66** (2019), no. 2, 221–226. MR3840124

[5] Mathematicians of the African Diaspora. `http://www.math.buffalo.edu/mad`.

[6] NAM Newsletter Archive. `https://www.nam-math.org/archives.html`.

602 West Main Street, Elizabeth City, North Carolina 27909
Email address: `jlhouston602@gmail.com`

Contemporary Mathematics
Volume **759**, 2020
https://doi.org/10.1090/conm/759/15263

An overview of mathematical modeling of geometric optics problems involving refraction

Henok Mawi

ABSTRACT. The inverse problem of determining shapes of surfaces which could transport a beam of rays from a given point source with a given intensity onto a set of directions or onto a target set with capabilities of controlling the output intensity is a complex problem that has several applications in designing optical systems. The theoretical investigation of the problem interweaves techniques used in the mathematics of mass transportation theory, convex analysis, calculus of variations and nonlinear partial differential equations of Monge-Ampère type. The development of computational techniques for the problem is important for practical applications. Here, we focus on the refractor problem with point source and overview some recent results.

1. Introduction

In 1993, among the open problems in geometry that were stated by Yau in [**80**], part of problem 21 reads as: *"For each point in the unit sphere S^2 in space we can issue a ray from the origin. If the ray reflects according to geometric optics, the direction of the reflected ray defines a point on the sphere. Hence we obtain a map from S^2 into S^2. How much information does this map tell us about the surface? Can we get this information numerically efficiently?"* The geometric optics problem of designing surfaces which could transport a beam of rays from a given source with a given intensity onto a set of directions or onto a target set with capabilities of controlling the output intensity is one such problem. However, the investigation into such beam reshaping surfaces started decades before the aforementioned problem by Yau. There are preceding works in [**60, 61**] and also in the engineering literature in [**18, 79**]. to mention a few. The surfaces have numerous applications in the design of optical devices. It is used in laser optics, [**7, 15, 52**] in reflector antennas [**8, 67, 78**], in photovoltaics (*solar concentrators*), in several medical devices , for example in *Lithotripter*, [**55**] and other areas such as street lighting, mobile displays and automotive headlight.

Historically, spherical and conic shapes were used to steer a beam of light and achieve a desired illuminance distribution, [**70**]. This classical approach of using rotationally symmetric surfaces has the disadvantage of loss and it is also known

2010 *Mathematics Subject Classification.* 35J96, 78A05, 52A15.

Key words and phrases. Geometric Optical design, Refraction, Optimal transport, Monge-Ampère equation, Iterative method, Minkowski method, Generated Jacobian Equations.

The author is supported, in part, by NSF grant HRD –1700236.

that the use of optical surfaces or lenses with nonsymmetric features; known in the literature as *freeform optical surfaces,* is more efficient in controlling output energy distribution. See the survey article [**58**] and the references there. The notion of freeform optical surfaces is considered a revolution in optical design [**70**] and there is a lot of interest in the area. See for example [**3, 17, 20, 53, 68**].

If the design problem is with in one medium and the synthesis involves only the law of reflection, it will be called a *Reflector Problem*, where as if the design problem involves propagation in different media with different indices of refraction and the law of refraction is applied it is called a *Refractor Problem.* In either case the objective could be to create the desired illumination in the the *far field* or the *near field* . In the far field case the goal is to send radiation into a set of directions where as in the near field case it is to send radiation to a specific target set. The source of light could also be a punctual source located a point or an extended source.

Several studies have been done on different aspects of these problems, both by the mathematics and engineering community, to refine the optical design of surfaces and extend the range of surface gradients to obtain more efficient devices. The literature is vast, but to mention a few; mathematical theory and modeling of the reflector problem is studied in [**9, 10, 12, 24, 42, 46, 57, 59, 60, 67, 76, 78**], more references can also be found [**58**]. Analysis for the refractor problem could be found in [**27, 29, 31, 32, 36, 40, 53, 62, 63**] and the references therein.

The design problem is a complex inverse problem which is based on the systematic application of the laws of reflection / refraction in geometrical optics and energy conservation principles in order to transport a given intensity of light from a source in such a way that the intensity is distributed in a prescribed manner on to a target. The mathematical analysis of the problem leads to fully nonlinear partial differential equations (PDE) of Monge-Ampère type, see [**24**] or to the more general form of generated Jacobian equations [**25**] .

In some cases, as in the far field case, the problems can be cast in optimal mass transportation problem framework with logarithmic cost function as shown in [**21, 77**] for the far field reflector problem and in [**31**] for the far field refractor problem. In those cases it is possible to recover the surface and deduce regularity results by appealing to the associated partial differential equations of Monge-Ampère type which is satisfied by the potential function. See [**48, 72, 73**] and the book of Villani [**75**] which contains a comprehensive discussion on optimal transport. For most part, the focus has been on the far field problems. Recently, more emphasis is given to the near field problems. In these cases the problem doesn't have a variational formulation and it can't be cast as optimal mass transport problem. However, the corresponding nonlinear Monge-Ampère type of PDE which is a special case of the Generated Jacobian Equations, can be obtained by using energy conservation and by computing the Jacobi determinant of the ray tracing map as is done in [**24, 31, 40, 42, 61**]. In fact, the PDE that results from near field problem in optics is the main motivation behind the introduction of Generated Jacobian Equations by Trudinger, [**25**]. An alternative approach to the problems is using Minkowski method [**65**] by exploiting mostly the geometric features of the problem.

A lot of progress has been made concerning existence, uniqueness and regularity of solutions. To mention a few; existence of global reflector by using Minkowski method is proved in [**10**], existence of smooth reflectors is studied in [**76**] and [**24**].

Regularity of (far field) reflector antennas is considered by Caffarelli, Gutiérrez and Huang in [**12**], and by Karakhanyan and Wang in [**42**]. Existence of far field refractor for a point source is proved by Gutiérrez and Huang in [**31**] by using optimal mass transport formulation, and $C^{1,\alpha}$ regularity of refractor surface for radiations issued from an extended source are proved in [**36**] and [**2**]. Local C^2 regularity of the solution to the refractor problem is studied in [**40**] by studying the associated Monge-Ampère type PDE and using A3 condition. A problem involving two refractor system that could form a lens which provides a prescribed light distribution from an extended source is studied in [**33**] and [**34**]. In [**35**], the authors studied the refractor problem in which one of the media of propagation is a metamaterial; an artificial materials which has negative index of refraction and which can be used in cloaking, improving medical imaging and providing high resolutions in microscopes (see [**23**] and the references there in).

The design and construction of such freeform optical surfaces have several important applications in systems involving both imaging and nonimaging optics. For application purposes, it will be necessary to devise computational methods and implement them numerically to design and fabricate beam redirecting surfaces. A numerical scheme to approximate the far field reflector antenna is developed in [**11**] by Caffarelli et al. Castro et al [**13**] have introduced numerical methods to solve the reflector problem. To design refractor surfaces a numerical algorithm is designed in [**14**] for the far field problem and recently in [**30**] for the near field problem. An algorithm for near field reflector is studied in [**5, 56, 68**] and for the related optimal mass transport problems under certain structure on the cost functions can be found in [**43, 44**]. Algorithm for the more general problem involving generated Jacobian equations is studied in [**1**]. More computational methods in the engineering literature can be found in the references in [**3, 17, 20, 53, 58, 68**].

In this survey we focus mainly on the mathematical aspects of the refractor problem with point source. For the reflector problem the reader may refer to the comprehensive survey by Oliker in [**58**]. In this paper we first, for completeness, recap the Law of Refraction (*Snell's Law*). In section 3 we state the far field refractor problem and discuss analytical and numerical approaches to solve the problem. We also mention some results regarding regularity of the solution to the problem. In the last section we will pose the near field refractor problem, discuss solutions following [**31**] and state some recent regularity and numerical results. The problems are practical and the models are used in three dimensions; however, it should be noted that the results that are discussed here are obtained in N dimensions. We also assume throughout this survey article, unless stated otherwise, medium I is denser than medium II and therefore $\kappa < 1$.

2. Refraction

Suppose from a point source of monochromatic light located at the origin O in $\mathbb{R}^3$ and surrounded by a homogeneous isotropic media I, a ray of light propagating in the direction of the unit vector $x \in S^2$, the unit sphere in $\mathbb{R}^3$, hits an interface $\mathbf{\Gamma}$ between medium I and another homogenous and isotropic medium, medium II obliquely, at a point where the unit normal $\nu \in S^2$ to $\mathbf{\Gamma}$ pointing towards medium II exists. Then refraction or change of direction happens as it propagates in medium II. Let m be a unit vector in the direction of the refracted ray. If v_1 and v_2 are the respective speeds of the light in medium I and medium II, then according to the law of refraction (Snell's Law) $\dfrac{\sin\theta_1}{\sin\theta_2} = \dfrac{v_1}{v_2}$ where θ_1 is the angle between x and ν (*angle of incidence*) and θ_2 is the angle between m and ν (*angle of refraction*). In vector form the law is given as

$$x - \kappa m = \lambda\nu \tag{2.1}$$

where $\lambda = x\cdot\nu - \kappa\sqrt{1-\kappa^{-2}(1-(x\cdot\nu)^2)}$, $\kappa = v_1/v_2$.

When medium I is optically denser than medium II and consequently $v_1 < v_2$, or equivalently $\kappa < 1$, the refracted ray bends away from the normal. As a result, there is a critical value θ_c, of the angle of incidence given by $\sin\theta_c = \kappa$ for which the refracted ray emerges in a direction tangent to the boundary. If the angle of incidence is larger than the critical angle no light enters medium II. This phenomenon is called *total internal reflection.* It can then be verified that there is no refraction unless $x\cdot m \geq \kappa$. By the reversibility of light rays, a similar observation can be done for $\kappa > 1$. Therefore a light ray in medium I in the direction $x \in S^2$ is refracted by some surface into a light ray in medium II in the direction $m \in S^2$ if and only if $x\cdot m \geq \kappa$ when $k < 1$ and if and only if $x\cdot m \geq 1/\kappa$ when $\kappa > 1$. Detail information on the law of refraction in geometric optics can be found in [**4**] and [**45**].

3. The far field refractor problem with point source

Assume that we are given two domains Ω, Ω^* in S^2 and suppose from a point source of light at the origin O, surrounded by media I, a monochromatic light is issued in the direction x with intensity density function $g(x)$ for $x \in \Omega$ where $g \in L^1(\Omega)$ and $g > 0\, a.e.$ on Ω. Suppose we are also given a Radon measure μ on Ω^*. The pair (Ω, g) corresponds to directions of incident light along with the intensity of light from the source and the pair (Ω^*, μ) corresponds to the directions of refracted rays along with a desired intensity distribution on the target.

In the far field refractor problem we would like to recover a refracting interface surface $\mathcal{R}$ between media I and II, given by a radial function ρ as $\mathcal{R} = \{\rho(x)x : x \in \Omega\}$ such that all rays emitted from the point O with directions $x \in \Omega$ are refracted by the surface $\mathcal{R}$ into media II with directions in Ω^* and the illumination intensity received on Ω^* is dictated by a Radon measure μ prescribed on Ω^*. Here we assume that there is no loss of energy due to internal reflection and g and μ satisfy the mass balance condition

$$\mu(\Omega^*) = \int_\Omega g(x)\,dx \tag{3.1}$$

with dx being the area element on S^2. Since Ω and Ω^* correspond to incident and refracted rays we impose the geometric condition

$$\inf_{x\in\Omega,\, m\in\Omega^*} x\cdot m \geq \kappa$$

to guarantee that any ray in the direction $x \in \Omega$ can be refracted into any ray in the direction $m \in \Omega^*$. See Lemma 2.1 in [**32**].

To formulate the problem more precisely we first observe that if $\mathcal{R}$ is a smooth interface, the Snell's law in equation (2.1) induces a set valued map $T_{\mathcal{R}} : \Omega^* \to \Omega$ which assigns to $m \in \Omega^*$ the set $T_{\mathcal{R}}(m)$ of directions $x \in \Omega$ which refract off $\mathcal{R}$ in the direction of m; $T_{\mathcal{R}}$ is called the *tracing mapping.* It can be shown [**32**], that the collection $\{V \subset \Omega^* : T_{\mathcal{R}}(V)\text{ is Lebesgue Measurable}\}$ is a $\sigma-$ algebra containing all

Borel sets in Ω^*. Also for any $m \in \Omega^*$ and $b > 0$ the semi-ellipsoid of revolution with one focus at the origin O, eccentricity κ and second focus at $\dfrac{2\kappa b}{1-\kappa^2}m$ given in polar form as:

$$E(m,b) = \left\{ \frac{b}{1-\kappa m\cdot x}x : \; x \in S^2, \; m\cdot x \geq \kappa \right\} \tag{3.2}$$

has a uniform refraction property. That is, if $E(m,b)$ is an interface between media I and II, all rays emanating from O in a direction x with $x\cdot m \geq \kappa$ will be refracted off $E(m,b)$ in the direction m.

DEFINITION 3.1. A surface $\mathcal{R}$ in $\mathbb{R}^3$ parameterized by $\rho(x)x$ is a *refractor from* $\overline{\Omega}$ *to* $\overline{\Omega^*}$ if for any $x_o \in \overline{\Omega}$ there exists a semi-ellipsoid $E(m,b)$ with $m \in \overline{\Omega^*}$ such that $\rho(x_o) = \dfrac{b}{1-\kappa m\cdot x_o}$ and $\rho(x) \leq \dfrac{b}{1-\kappa m\cdot x}$ for all $x \in \overline{\Omega}$. We call $E(m,b)$ a supporting semi-ellipsoid to $\mathcal{R}$ at $\rho(x_o)x_o$ or simply at x_o.

From the definition, it is not difficult to see that refractors are Lipschitz continuous in $\overline{\Omega}$, and therefore they are differentiable a.e. in Ω. Thus the mapping $T_{\mathcal{R}} : \Omega^* \to \Omega$ is defined for a. e. $m \in \Omega^*$.

A refractor $\mathcal{R}$ from $\overline{\Omega}$ to $\overline{\Omega^*}$ given as $\mathcal{R} = \{\rho(x)x : x \in \Omega\}$ is a (weak) *solution to the far field refractor problem* from (Ω, g) to (Ω^*, μ) if:

$$\begin{cases} \displaystyle\int_{T_{\mathcal{R}}(V)} g(x)\,dx = \mu(V) \\ \\ T_{\mathcal{R}}(\Omega^*) = \Omega. \end{cases} \tag{3.3}$$

for every Borel subset V of Ω^*. See an illustration in Fig. 1 below where $d\mu = f\,dx$. From Remark 5.6 in [**29**] a solution of the refractor problem is invariant under dilations by a constant. That is, if $\mathcal{R} = \{\rho(x)x : x \in \Omega\}$ is a solution and $c > 0$ then, so is $\mathcal{R}_c = \{c\rho(x)x : x \in \Omega\}$. Therefore with out loss of generality we can normalize a refractor $\mathcal{R} = \{\rho(x)x : x \in \overline{\Omega}\}$ from $\overline{\Omega}$ to $\overline{\Omega^*}$ so that it satisfies $\inf_{x\in\overline{\Omega}} \rho(x) = 1$. Then, by using the supporting semi-ellipsoids it is deduced that there is a uniform bound C such that for any normailized refractor,

$$\sup_{x\in\overline{\Omega}} \rho(x) \leq C. \tag{3.4}$$

The existence and uniqueness up to dilations of solution for the far field refractor problem was studied in [**32**] by casting it as an optimal mass transport problem with a logarithmic cost function. By using Minkowski method, existence of solutions was studied in [**29**] for the more general case where energy loss due to internal reflection is considered. A rigorous computational approach was developed to solve the problem to arbitrary precision in [**14**] where the convergence of the scheme is also proved. In the following subsections we will outline some of these results.

3.1. Existence of solutions. In this subsection we discuss the method of approximation by semi-ellipsoids to obtain existence of solutions to the far field refractor problem. By using Minkowski like method [**65**], the following result is obtained in [**29**].

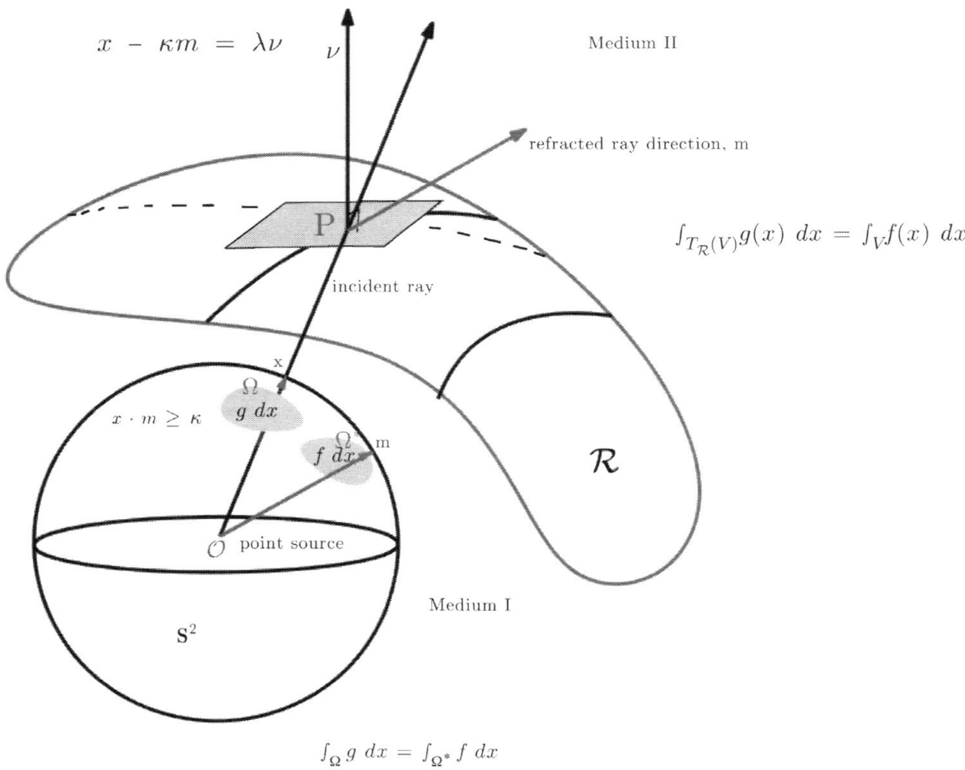

FIGURE 1. Far Field Refractor Problem

THEOREM 3.2. *Let $g \in L^1(\overline{\Omega})$ with $g > 0$, and let μ be a Radon measure on $\overline{\Omega^*}$, that satisfy* (3.1) *There exists a solution $\mathcal{R}$ of the refractor problem with emitting illumination intensity g and prescribed refracted illumination intensity μ.*

We will briefly discuss the the proof of the theorem. The idea is to prove the theorem first when μ is a discrete measure and use approximation by discrete measures when μ is a Radon measure. To that end suppose $m_1, m_2, \ldots, m_N$, $N \geq 2$ are distinct points in Ω^*. Let $f_1, \ldots, f_k$ be positive numbers and μ be a linear combination of dirac measures given as $\mu = \sum_{i=1}^k f_i \delta_i$ where δ_i is a dirac measure concentrated at m_i, for $1 \leq i \leq N$. Then for any $\mathbf{b} = (b_1, \ldots, b_N) \in \mathbb{R}^N$ with each $b_i > 0$ the multifaceted surface obtained from the semi-ellipsoids $E(m_i, b_i)$ for $1 \leq i \leq N$ defined by $\mathcal{R}(\mathbf{b}) = \{\rho(x)x : x \in \Omega\}$, where

$$\rho(x) = \min_{1 \leq i \leq N} \frac{b_i}{1 - \kappa m_i \cdot x}$$

is a refractor from Ω to Ω^*. It is not difficult to see that $T_{\mathcal{R}(\mathbf{b})}(\Omega^*) = \Omega$. Also, for fixed $m_1, m_2, \ldots, m_N$, the energy redistribution by $\mathcal{R}(\mathbf{b})$ on to Ω^* is completely determined by $\mathbf{b}$. It thus remains to find a vector $\mathbf{b} = (b_1, \ldots, b_N)$ so that the right amount of energy is sent in the direction m_i. That is:

$$G_i(\mathbf{b}) := \int_{T_{\mathcal{R}(\mathbf{b})}(m_i)} g(x)\, dx = f_i \tag{3.5}$$

for $i = 1, 2, \dots, N$. In the sequel, $\int_{T_{\mathcal{R}(\mathbf{b})}(m_i)} g(x)\,dx$ will be denoted by $G_i(\mathbf{b})$ or by $G_{\mathcal{R}(\mathbf{b})}(m_i)$ as relevant.

To show the existence of such a $\mathbf{b}$, first fix $b_1 = 1$ and let $W \subset \mathbb{R}^N$ be given by

$$W = \{\mathbf{b} = (1, b_2, \dots, b_N) : b_i > 0, \quad G_{\mathcal{R}(\mathbf{b})}(m_i) \le f_i \text{ for } i = 2, \dots, N\}.$$

Notice that if we increase b_i, keeping b_j unchanged for $j \ne i$, then the set $T_{\mathcal{R}(\mathbf{b})}(m_i)$ gets smaller and thus $G_i(\mathbf{b})$ decreases. It can also be seen from Lemma 3.6 in [**14**] that $G_i(\mathbf{b})$ is a continuous function of $\mathbf{b} \in W$ and

$$\lim_{b_i \to 0} G_i(\mathbf{b}) = \mu(\Omega^*) \quad \text{and} \quad \lim_{b_i \to \infty} G_i(\mathbf{b}) = 0. \tag{3.6}$$

Thus by taking b_i to be large enough for $i = 2, \dots, N$ we can show that $W \ne \emptyset$. Since the singular set of $\mathcal{R}(\mathbf{b})$ has measure zero, we have that

$$\sum_{i=1}^{N} \int_{T_{\mathcal{R}(\mathbf{b})}(m_i)} g(x)\,dx = \mu(\Omega^*) = \sum_{i=1}^{N} f_i.$$

By using this fact it can deduced that for $\mathbf{b} = (1, b_2, \dots, b_N) \in W$, $|T_{\mathcal{R}(\mathbf{b})}(m_1)| > 0$ and consequently $\frac{1}{1+\kappa} \le b_i$ for $i = 2, \dots, N$. Therefore given a fixed $\mathbf{b}^o = (1, b_2^o, \dots, b_N^o) \in W$ the set $W^o = \{\mathbf{b} = (1, b_2, \dots, b_N) \in W : b_i \le b_i^o\}$ is compact. The closedness is guaranteed by the continuity of $G_i(\mathbf{b})$, [**64**]. Consider now the function $\phi(\mathbf{b}) = \sum_{i=1}^{N} b_i$ on W^o. ϕ is continuous function and attains an absolute minimizer $\mathbf{b}^*$ on W^o. Then $\mathcal{R}(\mathbf{b}^*)$ is the refractor sought in Theorem 3.2 when μ is a discrete the measure given by $\mu = \sum_{i=1}^{k} f_i \delta_i$. Indeed otherwise, assume that $G_{\mathcal{R}(\mathbf{b}^*)}(m_2) < f_2$ and let $\mathcal{R}(\mathbf{b}_\lambda^*) = \{\rho_\lambda^*(x)x : x \in \overline{\Omega}\}$, where $\mathbf{b}_\lambda^* = (1, \lambda b_2^*, \dots, b_k^*)$ and $0 < \lambda < 1$. From the monotonocity property of G_i

$$G_{\mathcal{R}(\mathbf{b}_\lambda^*)}(m_i) \le G_{\mathcal{R}(\mathbf{b}^*)}(m_i) \le f_i \quad \text{for} \quad i \ne 2.$$

From the continuity of G_i we can choose λ sufficiently close to 1 so that $G_{\mathcal{R}(\mathbf{b}_\lambda^*)}(m_2) < f_2$. But then $\mathbf{b}_\lambda^* \in W_o \subset W$ with $\phi(\mathbf{b}_\lambda^*) < \phi(\mathbf{b}^*)$ which is a contradiction.

For the general case, for a given radon measure μ we first partition $\overline{\Omega^*}$ in to a finite number of disjoint Borel subsets $\omega_1^l, \dots, \omega_{k_l}^l$ such that $diam(\omega_i^l) \le \dfrac{1}{l}$ and obtain a sequence $\mathcal{R}_\ell := \mathcal{R}(\mathbf{b}_\ell) = \{\rho_\ell(x)x : x \in \overline{\Omega}\}$ of refractors corresponding to the discrete measure $\mu_\ell = \sum_{i=1}^{k_\ell} \mu(\omega_i^\ell)\delta_{m_i^\ell}$ where $m_i^\ell \in w_i^\ell$. By normalizing ρ_ℓ so that $\inf_{x\in\overline{\Omega}} \rho_\ell(x) = 1$ and by using equation (3.4) we obtain a uniformly bounded, equicontinuous family $\mathcal{R}_\ell = \{\rho_l : l \ge 1\}$ of refractors and by using *Arzelà - Ascoli*, if need be by taking subsequence, we conclude that $\rho_l \to \rho^*$ uniformly on $\overline{\Omega}$. Then $\mathcal{R}^* = \{\rho^*(x)x : x \in \overline{\Omega}\}$ is the refractror of Theorem 3.2 when μ is a general measure.

3.2. Uniqueness and regularity. An alternative approach by using optimal mass transport frame work [**74**] with logarithmic cost function

$$\log \frac{1}{1 - \kappa x \cdot m}$$

is carried out in [**32**] to show existence of solution to the far field refractor problem. In the same work, under the additional assumption that $\int_G g(x)\,dx > 0$ for any open set $G \subset \Omega$ it is also shown that the solution is unique up to dilation by a constant. That is if $\mathcal{R}_i = \{\rho_i(x)x : x \in \overline{\Omega}\}$, $i = 1, 2$ are two weak solutions to problem 3.3, then $\rho_1 = c\,\rho_2$ for some $c > 0$. In the case μ is a discrete measure and that the refractor solution is determined by finitely many semi ellipsoids, the uniqueness up to dilation is proved in [**27**] as well.

To study the regularity, first step is to derive the Monge-Ampère equation satisfied by ρ. To that assume $d\mu = f\,dx$. Then the mass balance equation 3.1 will take the form:

$$\int_{\Omega} g(x)\,dx = \int_{\Omega^*} f(x)\,dx.$$

If $\mathcal{R} = \{\rho(x)x : x \in \overline{\Omega}\}$ is a solution to the refractor problem 3.3 from (Ω, g) to (Ω^*, f) and $T : \Omega \to \Omega^*$ is defined by the relation (2.1) as

$$T(x) = m := \frac{1}{\kappa}(x - \lambda\nu) \tag{3.7}$$

the determinant of the Jacobian of T can be computed as in [**29**] to be

$$|\det DT| = \frac{dS_{\overline{\Omega^*}}}{dS_{\mathcal{U}}} = \frac{g(x')}{\sqrt{1-|x'|^2} f(T(x'))} \tag{3.8}$$

where $\mathcal{U} = \{x' = (x_1, x_2) : (x', \sqrt{1-|x'|^2}) \in \overline{\Omega}\}$ is the orthogonal projection of Ω, dS_Ω and $dS_{\mathcal{U}}$ denote the area elements corresponding to Ω and the volume element corresponding to $\mathcal{U}$ respectively.

If $\rho \in C^2(\overline{\Omega})$ then as in [**29**] the unit normal ν can be explicitly computed in terms of $D\rho$. Then equation (3.7) with (3.8) result in the Monge-Ampère type equation satisfied by ρ given as:

$$|Det(D^2\rho + A(x, \rho, D\rho))| = \psi(x, f, g, \rho, D\rho).$$

This is one way of obtaining the PDE satisfied by ρ. Alternatively, the far field refractor problem is a mass transportation problem and the associated potential function is $\phi = \log \rho$. The potential function then satisfies a fully nonlinear equation of Monge-Ampère type [**16**].

The regularity of the refractor is then analyzed by using the Ma-Trudinger-Wang (A3) condition introduced in [**48**]. For $\kappa < 1$ the **A3** condition doesn't hold [**32**] and therefore one can't expect C^1 regularity even for smooth densities. See the counterexample given [**51**]. For the case $\kappa > 1$ the following regularity result was deduced in [**39**] for smooth densities f and g based on the regularity result of potential functions for general cost functions.

THEOREM 3.3. *Suppose f and g satisfy $C_1 \le f, g \le C_2$ for some positive constants C_1 and C_2. If $k > 1$, the solutions are locally (globally) smooth provided f and g are smooth functions.*

3.3. Numerical approach to the far field problem. In this section we outline a numerical technique established in [**14**] for the construction of a surface that solves the far field refractor problem when μ is a discrete measure. The idea of the technique was originally applied in [**11**] to deal with the reflector problem. However, a major advance and simplification to solve these kind of problems numerically is introduced in [**14**] and [**1**] by showing that an appropriate mapping satisfies a Lipschitz condition. This essential step guarantees that the algorithm converges in a finite number of iterations without imposing a smoothness condition on the density function g. The technique was also extended in [**43**] for mass transport problems with cost functions satisfying the **A3** conditions given in [**48**], and more recently it is used to approximate a solution to the near field refractor case in [**30**].

The main idea of the numerical algorithm when μ is a discrete measure is as follows [**14**]. We fix our error $\epsilon > 0$, $m_1, m_2, \ldots, m_N$, $N \geq 2$ distinct points in Ω^* and $f_1, \ldots, f_N$ positive numbers. As before we let μ be a linear combination of dirac measures given as $\mu = \sum_{i=1}^N f_i \delta_i$ where δ_i is a dirac measure concentrated at m_i, for $1 \leq i \leq N$. Let $\mathbb{R}_+^N = \{\mathbf{b} = (b_1, \ldots, b_N) : b_i > 0 \text{ for all } i = 1, \ldots, N\}$. Take $\delta < \epsilon/N$ and set

$$W^\delta = \{\mathbf{b} \in \mathbb{R}_+^N : b_1 = 1, G_{\mathcal{R}(\mathbf{b})}(m_i) < f_i + \delta, \forall i = 2, \ldots, N\}$$

The iterative scheme is to obtain a $\mathbf{b}^* \in W^\delta$ so that $|G_i(\mathbf{b}) - f_i| < \epsilon$ for all i.

From equation (3.6) it is possible to see that $W^\delta \neq \emptyset$ and also that by a similar argument as in Section 3.1, $\frac{1}{1+\kappa} \leq b_i$ for all $1 \leq i \leq N$. We Pick $\mathbf{b}^1 = (1, b_2, \cdots, b_N) \in W^\delta$ and initialize the algorithm. If $|G_{\mathcal{R}(\mathbf{b}^1)}(m_i) - f_i| < \delta$ for $i = 2, \ldots, N$ then $\mathbf{b}^1 = \mathbf{b}^\epsilon$. Indeed from the (3.1)

$$|G_{\mathcal{R}(\mathbf{b}^1)}(m_i) - f_i| < \delta < \epsilon \quad \text{for } 2 \leq i \leq N$$

and

$$\begin{aligned} |f_1 - G_{\mathcal{R}(\mathbf{b}^1)}(m_1)| &= \left| \sum_{i=2}^N G_{\mathcal{R}(\mathbf{b}^1)}(m_i) - f_i \right| \\ &\leq \sum_{i=2}^N |G_{\mathcal{R}(\mathbf{b}^1)}(m_i) - f_i| \leq (N-1)\,\delta < \epsilon \end{aligned}$$

Otherwise set $\mathbf{b}^{1,1} = \mathbf{b}^1$ and construct $N-1$ intermediate vectors $\mathbf{b}^{1,2}, \ldots, \mathbf{b}^{1,N}$ in W^δ to obtain $\mathbf{b}^2$ as follows. While $1 \leq i \leq N-1$, given $\mathbf{b}^{1,i} \in W^\delta$,

Step 1: Test if $|G_{\mathcal{R}(\mathbf{b}^{1,i})}(m_{i+1}) - f_{i+1}| < \delta$.

Case 1: if True, set $\mathbf{b}^{1,i+1} = \mathbf{b}^{1,i}$ and test if $|G_{\mathcal{R}(\mathbf{b}^{1,i+1})}(m_{i+2}) - f_{i+2}| < \delta$.

Case 2: else

$$G_{\mathcal{R}(\mathbf{b}^{1,i})}(m_{i+1}) > f_{i+1} + \delta$$

or

$$G_{\mathcal{R}(\mathbf{b}^{1,i})}(m_{i+1}) < f_{i+1} - \delta.$$

Since $\mathbf{b}^{1,i} \in W^\delta$, $G_{\mathcal{R}(\mathbf{b}^{1,i})}(m_{i+1}) > f_{i+1} + \delta$ is not possible. By using the continuity and monotonocity of G_i pick $b_{i+1}^* \in (0, b_{i+1}^{1,i})$ so that $\mathbf{b}^{1,i+1} := (1, b_2, \ldots, b_i, b_{i+1}^*, \ldots, b_N)$ satisfies $f_{i+1} < G_{\mathcal{R}(\mathbf{b}^{1,i+1})}(m_{i+1}) < f_{i+1} + \delta$. Note $\mathbf{b}^{1,i+1} \in W^\delta$.

Step 2: Repeat Step 1 with $\mathbf{b}^{1,i+1} := (1, b_2, \ldots, b_i, b_{i+1}^*, \ldots, b_N)$

Step 3: Set $\mathbf{b}^2 = \mathbf{b}^{2,1} = \mathbf{b}^{1,N}$ and repeat.

Notice that in this iteration if $\mathbf{b}^{n,j-1} = (1, b_2^{n,j-1}, \ldots, b_N^{n,j-1})$ and $\mathbf{b}^{n,j} = (1, b_2^{n,j}, \ldots, b_N^{n,j})$ are two successive vectors in the iteration and $\mathbf{b}^{n,j-1} \neq \mathbf{b}^{n,j}$ then

$$b_j^{n,j} < b_j^{n,j-1}$$

and by monotonocity of G_j

$$G_j(\mathbf{b}^{n,j-1}) < f_j - \delta < f_j < G_j(\mathbf{b}^{n,j}). \tag{3.9}$$

Also, for any iterate $\mathbf{b}^{n,\ell} = (1, b_2^{n,\ell}, \dots, b_N^{n,\ell})$, $\mathbf{b}^{n,\ell} \in W^\delta$ and

$$\frac{1}{1+\kappa} < b_j^{n,\ell} < b_j^1$$

So, we obtain a bounded monotone sequence of vectors $\{\mathbf{b}^j\} \in W^\delta$ and corresponding refractors $\mathcal{R}(\mathbf{b}^j)$. If for some n_o, $\mathbf{b}^{n_o} = \mathbf{b}^{n_o+1}$ it means that none of the entries of the vector $\mathbf{b}^n$ were changed in the intermediate steps, which in turn means that the test in **Step 1** has come true for all $1 \le i \le N-1$. Thus $|G_{\mathcal{R}(\mathbf{b}^n)}(m_i) - f_i| < \delta$ for $i = 2, \dots, N$ and as in above $\mathbf{b}^\epsilon = \mathbf{b}^n = \mathbf{b}^{n+1}$. The existence of such an n_o and consequently an approximate solution $\mathcal{R}(\mathbf{b}^{n_o})$ is guaranteed after proving that G_i is Lipschitz in b_i. The Lipschitz constant along with the relation in (3.9) imposes a lower bound on the step size decrease of the ith entry during iteration there by imposing an upper bound on the number of iterations. See [**14**] for details.

To conclude this section we make some remarks related to the far field refractor problem.

1. The procedure described in this subsection 3.3 can also be used as a constructive way of obtaining the exact solution to the far field refractor problem when μ is discrete. Indeed, by the procedure, for each ϵ we obtain a b^ϵ and a corresponding $\mathcal{R}(\mathbf{b}^\epsilon)$ such that for $i = 1, \dots, N$,

$$|G_{\mathcal{R}(\mathbf{b}^\epsilon)}(m_i) - f_i| < \epsilon$$

By using the continuity of $G_{\mathcal{R}(\mathbf{b})}(m_i)$ and letting $\epsilon \to 0$ it is possible to obtain a refractor which satisfies 3.5.

2. Reflection and refraction of light happen in tandem. That is, if a ray of light hits a boundary $\mathcal{R}$ between the media I and II which have different optical properties, the ray is split between two rays; a refracted (transmitted) ray proceeding into medium II and a reflected wave propagated back in to medium I. As a result there will be loss of intensity due to reflection. A model problem of constructing an interface between media I and II in such a way that both the incident ray and the refracted ray are prescribed and the loss of energy due to internal reflection is taken into account is developed in [**29**]. However, it is not clear if there is a variational formulation as an optimal mass transport in the case when the energy loss due to internal reflection is taken into account.

3. When the media of propagation is anisotorpic; where the optical property at a point depends on direction, then the problem becomes more involved as rays may be refracted into two rays in medium II: an ordinary ray and an extraordinary ray. This is the phenomenon of bi-refringence, observed experimentally in crystals, see [**45**] for details. A model for the far field problem is suggested in [**28**] when both media I and II are anisotropic. There is no such model for the near field refractor problem.

4. The near field refractor problem with point source

Unlike the far field refractor problem where the output domain Ω^* is a set of directions in to which the refracted rays should propagate, in the near field refractor problem the output domain Ω^* is a prespecified target set in space usually contained in a hypersurface. The near field refractor problem is also different from the far field problem in that it doesn't have a mass transport formulation. Additionally, the geometry of the Cartesian Ovals; the building blocks to construct the surface solutions, is more difficult to handle. One of the earlier studies of the near field refractor problem is in [**53**] where μ is considered to be a discrete measure. The existence of solutions for more general measures is obtained in [**31**] for both cases when $\kappa < 1$ and $\kappa > 1$. In what follows we will state the near field refractor problem and briefly discuss existence of solutions.

4.1. Statement of the problem. To formulate the near field refractor problem with point source, consider a source of light located at the origin O shining through an aperture $\Omega \subset S^2$ with intensity density $g(x)$ for $x \in \Omega$. Let Ω^* be a region in $\mathbb{R}^3$ contained in some hypersurface, usually a plane and μ be a Radon measure on Ω^* satisfying the energy conservation $\mu(\Omega^*) = \int_\Omega g(x)\,dx$. We assume that Ω is surrounded by Media I and Ω^* is surrounded by media II. Some structural conditions are also imposed on Ω and Ω^* to avoid total internal reflection and obstruction of Ω^*, [**31**]. A (weak) solution to the *near field refractor problem* is then to find an interface $\mathcal{S}$ between media I and II parametrized by $\mathcal{S} = \{\rho(x)x : x \in \overline{\Omega}\}$ so that each ray with direction $x \in \Omega$ refracts into a point in Ω^* according to Snell's law and so that the energy conservation condition

$$\mathcal{M}_{\mathcal{S},f}(F) := \int_{\mathcal{T}_\mathcal{S}(F)} g(x)\,dx = \mu(F) \tag{4.1}$$

holds for all Borel sets $F \subset \Omega^*$, where $\mathcal{T}_\mathcal{S}(F)$ represents the directions $x \in \Omega$ issued from the origin O and that hit a point in F after being refracted by the surface $\mathcal{S}$. The set $\mathcal{T}_\mathcal{S}(F)$ is Lebesgue measurable and $\mathcal{M}_{\mathcal{S},f}(F)$ is a Radon measure on $\overline{\Omega}$. It is called the *refractor measure.*

An illustration of the near field refractor problem is shown in Figure 2, once again with $d\mu = f\ dx$.

4.2. Existence of solutions. For the far field refractor problem with $\kappa < 1$ we have seen that segments of semi-ellipsoids are used as building blocks to create the illumination pattern desired. For the near field problems the corresponding surfaces are Cartesian Ovals. Michaelis *et al.* in [**53**] have observed this by using Hamiltonian theory of ray optics and they used ideas developed for reflector in [**58**] to formulate and study the near field refractor problem . Their results were further elaborated in [**68**] and numerically implemented. In [**31**], Gutierrez and Huang have used Snell's law to deduce that the Cartesian oval $\mathcal{O}_b(P)$ given in polar coordinates as:

$$\mathcal{O}_b(P) = \{h(x,P,b)x : x \in S^{n-1},\ x \cdot P \geq b\}$$

where

$$h(x,P,b) = \frac{(b-\kappa^2 x\cdot P) - \sqrt{(b-\kappa^2 x\cdot P)^2 - (1-\kappa^2)(b^2-\kappa^2|P|^2)}}{1-\kappa^2} \tag{4.2}$$

is an optical surface that refracts light from a point source O located in medium I with refractive index n_1 to a single target point P located in medium II with refractive index n_2. See Figure 3. In this section we outline the work in [**31**] where the authors have formulated and studied existence of solutions for the near field refractor problem.

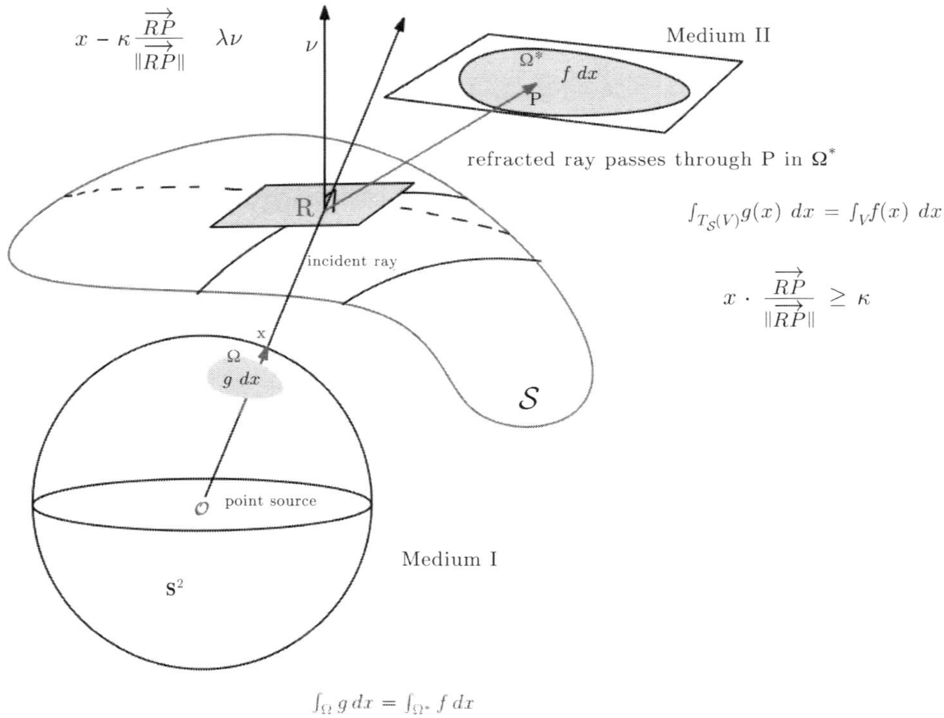

FIGURE 2. Near Field Refractor Problem

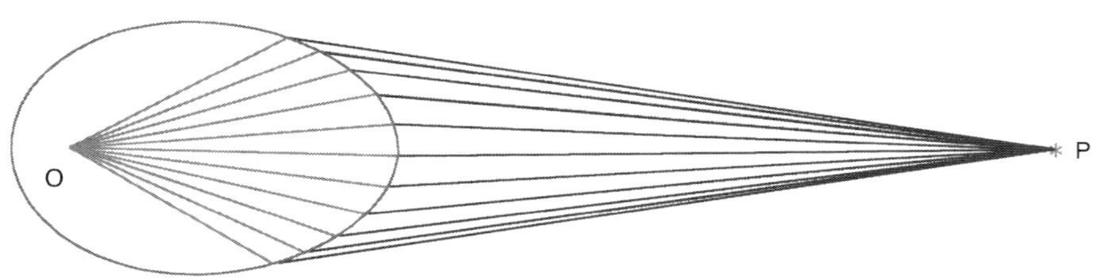

FIGURE 3. Refraction Property of Oval; $\kappa = 0.7$, $P = (10, 0)$, $b = 8$

A near field refractor is defined as follows:

DEFINITION 4.1. A surface $\mathcal{S} = \{x\rho(x) : x \in \overline{\Omega}\}$ is said to be a near field refractor if for any point $y\rho(y) \in \mathcal{S}$ there exist point $P \in \Omega^*$ and $\kappa|P| < b < |P|$ such that the refracting oval $\mathcal{O}_b(P)$ supports $\mathcal{S}$ at $y\rho(y)$, i.e. $\rho(x) \leq h(x, P, b)$ for all $x \in \overline{\Omega}$ with equality at $x = y$.

Notice that if the target set Ω^* is very far (i.e. $|P| \to \infty$), then ovals look more like ellipsoids and the near field problem approaches the far field problem.

To solve the problem, first, the case when μ is a linear combination of delta measures is considered. That is, given distinct points $P_1, \cdots, P_N$, in Ω^*, positive numbers $g_1, \cdots, g_N$, $\mu = \sum_{i=1}^{N} g_i\, \delta_{P_i}$ satisfying the energy conservation

$$\int_{\overline{\Omega}} f(x)dx = \sum_{i=1}^{N} g_i, \tag{4.3}$$

a point $X_o \in \Omega^*$ and $\kappa|P_1| < b_1 < \kappa|P_1| + r_0 \dfrac{(1-\kappa)^2}{1+\kappa}$, ($r_0$ is a constant that is related to the structure of Ω and Ω^*), it is shown that there is a unique vector $\mathbf{b} = (b_1, b_2, \ldots, b_N) \in \prod_{i=1}^{N}(\kappa|P_i|, |P_i|)$ such that the surface made of segments of the confocal cartesian ovals $\mathcal{O}_{b_i}(P_i)$ and given by

$$\mathcal{S}(\mathbf{b}) := \{\rho(x)x : x \in \overline{\Omega} \,\text{and}\, \rho(x) = \min_{1\le i\le N} h(x, P_i, b_i)\} \tag{4.4}$$

is a solution to the near field refractor problem passing through X_o. In particular it is shown that

$$\mathcal{M}_{\mathcal{S}(\mathbf{b}),f}(P_i) = g_i$$

for all $i = 1, \ldots, N$ and consequently from the additivity of the refractor measure along with the fact that $|\mathcal{T}_{\mathcal{S}}(P_i) \cap \mathcal{T}_{\mathcal{S}}(P_j)| = 0$, for $i \neq j$(4.1) holds.

For the general case the main idea of the proof is similar to the case for the far field refractor problem. Firstly, partition $\Omega^* = \cup_{i=1}^{N_\ell}\omega_i$ and for each $\ell \in \mathbb{N}$ define a measure $\mu_\ell = \sum_{i=1}^{n_\ell} \mu(\omega_i)\delta_{P_i}$ which is a linear combination of Dirac measures whose masses are points chosen from the partition, so that μ_ℓ converges weakly to μ. Clearly $\mu_\ell(\Omega^*) = \mu(\Omega^*)$. Secondly, from the above, obtain $\mathbf{b}_\ell$ and a near field refractor $\mathcal{S}(\mathbf{b}_\ell) = \{\rho_\ell(x)x : x \in \overline{\Omega}\}$ corresponding to the measure μ_ℓ and passing through a fixed point X_o. Finally, it is shown that the sequence $\{\rho_\ell : \ell \in \mathbb{N}\}$ is compact and the solution is obtained as a limit of the sequence. The existence result for the general case is stated as follows in [**31**].

THEOREM 4.2. *Let μ be a Radon measure on Ω^*, $f \in L^1(\Omega)$ with $f > 0$ a.e., and satisfying the energy conservation condition $\mu(\Omega^*) = \int_\Omega g(x)\,dx$. Then there exists a* weak solution *of the near field refractor problem; that is there is a refractor surface $\mathcal{S}$ satisfying* (4.1) .

The Lipschitz continuity of a near field refractor with a Lipschitz constant depending on κ, $\max_{\Omega^*}|P|$ and the geometry of Ω and Ω^* is established in [**31**]. Further regularity properties of the solution to the near field refractor problem are obtained recently. In [**41**] the problem is cast in the framework of generated Jacobian equations, the corresponding nonlinear PDE is derived and local C^2 regularity of the solution to the near field refractor problem is obtained under smoothness assumptions on the density functions. In [**37**], $C^{1,\alpha}$ estimates are derived under structural conditions on the source and target set without assuming smoothness on the data. Both results are for $\kappa > 1$. Recently, an iterative algorithm to approximate the solution to the near field refractor problem when μ is a discrete measure is demonstrated and the convergence of the scheme is rigorously proved in [**30**].

References

[1] Farhan Abedin and Cristian E. Gutiérrez, *An iterative method for generated Jacobian equations*, Calc. Var. Partial Differential Equations **56** (2017), no. 4, Art. 101, 14, DOI 10.1007/s00526-017-1200-2. MR3669141

[2] Farhan Abedin, Cristian E. Gutiérrez, and Giulio Tralli, *$C^{1,\alpha}$ estimates for the parallel refractor*, Nonlinear Anal. **142** (2016), 1–25, DOI 10.1016/j.na.2016.04.001. MR3508055

[3] A. Bauer and J. P. Rolland, *Visual space assessment of two all reflective, free-form, optical see-through head-worn displays,* Opt. Express 22, 13155–13163 (2014).

[4] Born, M., Wolf, E. *Principles of Optics*, 7th ed., Cambridge University Press, 1999.

[5] Kolja Brix, Yasemin Hafizogullari, and Andreas Platen, *Solving the Monge-Ampère equations for the inverse reflector problem*, Math. Models Methods Appl. Sci. **25** (2015), no. 5, 803–837, DOI 10.1142/S0218202515500190. MR3319337

[6] Ioan Bucataru and Michael A. Slawinski, *Generalized orthogonality between rays and wavefronts in anisotropic inhomogeneous media*, Nonlinear Anal. Real World Appl. **6** (2005), no. 1, 111–121, DOI 10.1016/j.nonrwa.2004.03.004. MR2104545

[7] D.G. Burkhard and G.L. Strobel. Reflective optics system for multiple beam laser fusion, Applied Optics, 22(9):1313–1317, 1983.

[8] D. G. Burkhard and D. L. Shealy, Specular aspheric surface to obtain a specified irradiance from discrete or continuous line source radiation: design. Applied Optics, 14(6):1279–1284, 1975.

[9] Luis A. Caffarelli and Qingbo Huang, *Reflector problem in $\mathbb{R}^n$ endowed with non-Euclidean norm*, Arch. Ration. Mech. Anal. **193** (2009), no. 2, 445–473, DOI 10.1007/s00205-009-0221-1. MR2525123

[10] L. A. Caffarelli and V. I. Oliker, *Weak solutions of one inverse problem in geometric optics*, J. Math. Sci. (N.Y.) **154** (2008), no. 1, 39–49, DOI 10.1007/s10958-008-9152-x. Problems in mathematical analysis. No. 37. MR2493255

[11] Luis A. Caffarelli, Sergey A. Kochengin, and Vladimir I. Oliker, *On the numerical solution of the problem of reflector design with given far-field scattering data*, Monge Ampère equation: applications to geometry and optimization (Deerfield Beach, FL, 1997), Contemp. Math., vol. 226, Amer. Math. Soc., Providence, RI, 1999, pp. 13–32, DOI 10.1090/conm/226/03233. MR1660740

[12] Luis A. Caffarelli, Cristian E. Gutiérrez, and Qingbo Huang, *On the regularity of reflector antennas*, Ann. of Math. (2) **167** (2008), no. 1, 299–323, DOI 10.4007/annals.2008.167.299. MR2373156

[13] Pedro Machado Manhães de Castro, Quentin Mérigot, and Boris Thibert, *Far-field reflector problem and intersection of paraboloids*, Numer. Math. **134** (2016), no. 2, 389–411, DOI 10.1007/s00211-015-0780-z. MR3537955

[14] Roberto De Leo, Cristian E. Gutiérrez, and Henok Mawi, *On the numerical solution of the far field refractor problem*, Nonlinear Anal. **157** (2017), 123–145, DOI 10.1016/j.na.2017.03.009. MR3645803

[15] F. Dickey, S. Holswade, D. Shealy, *Laser Beam Shaping Applications*, Taylor and Francis Group, 2005.

[16] Lawrence C. Evans, *Partial differential equations and Monge-Kantorovich mass transfer*, Current developments in mathematics, 1997 (Cambridge, MA), Int. Press, Boston, MA, 1999, pp. 65–126. MR1698853

[17] G. W. Forbes, *Characterizing the shape of free-form optics* Opt. Express 20, 2483–2499 (2012).

[18] V. Galindo, *Design of dual-reflector antennas with arbitrary phase and amplitude distributions*, IEEE Transactions on Antennas and Propagation, **12** 4, (1963).

[19] Wilfrid Gangbo and Vladimir Oliker, *Existence of optimal maps in the reflector-type problems*, ESAIM Control Optim. Calc. Var. **13** (2007), no. 1, 93–106, DOI 10.1051/cocv:2006017. MR2282103

[20] Gannon, C. and R. Liang. *Optimal geometry for off-axis, freeform illumination design,* Optical Engineering 57, no. 6 (2018)

[21] T. Glimm and V. Oliker, *Optical design of single reflector systems and the Monge-Kantorovich mass transfer problem*, J. Math. Sci. (N.Y.) **117** (2003), no. 3, 4096–4108, DOI 10.1023/A:1024856201493. Nonlinear problems and function theory. MR2027449

[22] Tobias Graf and Vladimir I. Oliker, *An optimal mass transport approach to the near-field reflector problem in optical design*, Inverse Problems **28** (2012), no. 2, 025001, 15, DOI 10.1088/0266-5611/28/2/025001. MR2876852

[23] Allan Greenleaf, Yaroslav Kurylev, Matti Lassas, and Gunther Uhlmann, *Invisibility and inverse problems*, Bull. Amer. Math. Soc. (N.S.) **46** (2009), no. 1, 55–97, DOI 10.1090/S0273-0979-08-01232-9. MR2457072

[24] Pengfei Guan and Xu-Jia Wang, *On a Monge-Ampère equation arising in geometric optics*, J. Differential Geom. **48** (1998), no. 2, 205–223. MR1630253

[25] Nestor Guillen and Jun Kitagawa, *Pointwise estimates and regularity in geometric optics and other generated Jacobian equations*, Comm. Pure Appl. Math. **70** (2017), no. 6, 1146–1220, DOI 10.1002/cpa.21691. MR3639322

[26] Cristian E. Gutiérrez, *The Monge-Ampère equation*, Progress in Nonlinear Differential Equations and their Applications, vol. 44, Birkhäuser Boston, Inc., Boston, MA, 2001. MR1829162

[27] C. E. Gutiérrez, Refraction problems in geometric optics, in *Fully nonlinear PDEs in real and complex geometry and optics*, 95–150, Lecture Notes in Math., 2087, Fond. CIME/CIME Found. Subser, Springer, Cham.

[28] Cristian E. Gutiérrez, Qingbo Huang, and Henok Mawi, *Refractors in anisotropic media associated with norms*, Nonlinear Anal. **188** (2019), 125–141, DOI 10.1016/j.na.2019.05.020. MR3961154

[29] Cristian E. Gutiérrez and Henok Mawi, *The refractor problem with loss of energy*, Nonlinear Anal. **82** (2013), 12–46, DOI 10.1016/j.na.2012.11.024. MR3020894

[30] Roberto De Leo, Cristian E. Gutiérrez, and Henok Mawi, *On the numerical solution of the far field refractor problem*, Nonlinear Anal. **157** (2017), 123–145, DOI 10.1016/j.na.2017.03.009. MR3645803

[31] Cristian E. Gutiérrez and Qingbo Huang, *The near field refractor*, Ann. Inst. H. Poincaré Anal. Non Linéaire **31** (2014), no. 4, 655–684, DOI 10.1016/j.anihpc.2013.07.001. MR3249808

[32] Cristian E. Gutiérrez and Qingbo Huang, *The refractor problem in reshaping light beams*, Arch. Ration. Mech. Anal. **193** (2009), no. 2, 423–443, DOI 10.1007/s00205-008-0165-x. MR2525122

[33] Cristian E. Gutiérrez and Ahmad Sabra, *Aspherical lens design and imaging*, SIAM J. Imaging Sci. **9** (2016), no. 1, 386–411, DOI 10.1137/15M1030807. MR3477314

[34] Cristian E. Gutiérrez and Ahmad Sabra, *Freeform lens design for scattering data with general radiant fields*, Arch. Ration. Mech. Anal. **228** (2018), no. 2, 341–399, DOI 10.1007/s00205-017-1196-y. MR3766979

[35] Gutiérrez, C., Stachura, E., *Uniform Refraction in Negative Refractive Index Materials*, arXiv:1501.03109, 2015.

[36] Cristian E. Gutiérrez and Federico Tournier, *Regularity for the near field parallel refractor and reflector problems*, Calc. Var. Partial Differential Equations **54** (2015), no. 1, 917–949, DOI 10.1007/s00526-014-0811-0. MR3385186

[37] Gutiérrez, C. E., and Tournier, F. *$C^{1,\alpha}$ estimates for the near field refractor,* https://arxiv.org/abs/1904.02475.

[38] Julien André, Dominique Attali, Quentin Mérigot, and Boris Thibert, *Far-field reflector problem under design constraints*, Internat. J. Comput. Geom. Appl. **25** (2015), no. 2, 143–162, DOI 10.1142/S0218195915500090. MR3371424

[39] Aram L. Karakhanyan, *On the regularity of weak solutions to refractor problem*, Armen. J. Math. **2** (2009), no. 1, 28–37. MR2485874

[40] Aram L. Karakhanyan, *An inverse problem for the refractive surfaces with parallel lighting*, SIAM J. Math. Anal. **48** (2016), no. 1, 740–784, DOI 10.1137/140964941. MR3463050

[41] A. Karakhanyan and A. Sabra, *Refractor surfaces determined by near-field data*, arXiv:1810.07094

[42] Aram Karakhanyan and Xu-Jia Wang, *On the reflector shape design*, J. Differential Geom. **84** (2010), no. 3, 561–610. MR2669365

[43] Jun Kitagawa, *An iterative scheme for solving the optimal transportation problem*, Calc. Var. Partial Differential Equations **51** (2014), no. 1-2, 243–263, DOI 10.1007/s00526-013-0673-x. MR3247388

[44] Jun Kitagawa, Quentin Mérigot, and Boris Thibert, *Convergence of a Newton algorithm for semi-discrete optimal transport*, J. Eur. Math. Soc. (JEMS) **21** (2019), no. 9, 2603–2651, DOI 10.4171/JEMS/889. MR3985609

[45] Morris Kline and Irvin W. Kay, *Electromagnetic theory and geometrical optics*, Pure and Applied Mathematics, Vol. XII, Interscience Publishers John Wiley & Sons, Inc. New York-London-Sydney, 1965. MR0180094

[46] Sergey A. Kochengin and Vladimir I. Oliker, *Determination of reflector surfaces from near-field scattering data. II. Numerical solution*, Numer. Math. **79** (1998), no. 4, 553–568, DOI 10.1007/s002110050351. MR1631685

[47] Sergey A. Kochengin and Vladimir I. Oliker, *Determination of reflector surfaces from near-field scattering data*, Inverse Problems **13** (1997), no. 2, 363–373, DOI 10.1088/0266-5611/13/2/011. MR1445924

[48] Xi-Nan Ma, Neil S. Trudinger, and Xu-Jia Wang, *Regularity of potential functions of the optimal transportation problem*, Arch. Ration. Mech. Anal. **177** (2005), no. 2, 151–183, DOI 10.1007/s00205-005-0362-9. MR2188047

[49] Grégoire Loeper and Cédric Villani, *Regularity of optimal transport in curved geometry: the nonfocal case*, Duke Math. J. **151** (2010), no. 3, 431–485, DOI 10.1215/00127094-2010-003. MR2605867

[50] Grégoire Loeper, *Regularity of optimal maps on the sphere: the quadratic cost and the reflector antenna*, Arch. Ration. Mech. Anal. **199** (2011), no. 1, 269–289, DOI 10.1007/s00205-010-0330-x. MR2754343

[51] Grégoire Loeper, *On the regularity of solutions of optimal transportation problems*, Acta Math. **202** (2009), no. 2, 241–283, DOI 10.1007/s11511-009-0037-8. MR2506751

[52] S. Mirov, V. Oliker, L.D. Prussner, D. Shealy, *Optical design of a two-mirror asymmetrical reshaping system and its application in superbroadband color center lasers*, SPIE Proceedings, **2263** (1994), 10-18.

[53] D. Michaelis, P. Schreiber and A. Bräuer , *Cartesian oval representation of freeform optics in illumination systems*, Optics Letters Vol. 36, Issue 6, pp. 918-920 (2011).

[54] Y. Mazur and A. Voznesenskaya, *Numerical Modeling of Freeform Refractive Surfaces Forming Line Radiation Pattern*, Procedia Computer Science101 , 2016 .

[55] Neisius, A. *et al* , *Improving the lens design and performance of a contemporary electromagnetic shock wave lithotripter*, Proceedings of the National Academy of Sciences, Vol. 111, **13**, E1167–E1175, 2014.

[56] Andreas Neubauer, *Design of 3D-reflectors for near field and far field problems*, Large-scale optimization with applications, Part I (Minneapolis, MN, 1995), IMA Vol. Math. Appl., vol. 92, Springer, New York, 1997, pp. 101–118, DOI 10.1007/978-1-4612-1962-0_6. MR1465685

[57] Vladimir Oliker, *Designing freeform lenses for intensity and phase control of coherent light with help from geometry and mass transport*, Arch. Ration. Mech. Anal. **201** (2011), no. 3, 1013–1045, DOI 10.1007/s00205-011-0419-x. MR2824470

[58] Vladimir Oliker, *Mathematical aspects of design of beam shaping surfaces in geometrical optics*, Trends in nonlinear analysis, Springer, Berlin, 2003, pp. 193–224. MR1999099

[59] Vladimir Oliker, *The reflector problem for closed surfaces*, Partial differential equations and applications, Lecture Notes in Pure and Appl. Math., vol. 177, Dekker, New York, 1996, pp. 265–270. MR1371597

[60] V. I. Oliker, *Near radially symmetric solutions of an inverse problem in geometric optics*, Inverse Problems **3** (1987), no. 4, 743–756. MR928051

[61] V. Oliker and P. Waltman, *Radially symmetric solutions of a Monge-Ampère equation arising in a reflector mapping problem*, Differential equations and mathematical physics (Birmingham, Ala., 1986), Lecture Notes in Math., vol. 1285, Springer, Berlin, 1987, pp. 361–374, DOI 10.1007/BFb0080616. MR921288

[62] Vladimir Oliker, *On design of free-form refractive beam shapers, sensitivity to figure error, and convexity of lenses*, J. Opt. Soc. Amer. A **25** (2008), no. 12, 3067–3076, DOI 10.1364/JOSAA.25.003067. MR2479632

[63] Vladimir Oliker, Jacob Rubinstein, and Gershon Wolansky, *Supporting quadric method in optical design of freeform lenses for illumination control of a collimated light*, Adv. in Appl. Math. **62** (2015), 160–183, DOI 10.1016/j.aam.2014.09.009. MR3277469

[64] A. V. Pogorelov, *Monge-Ampère equations of elliptic type*, Translated from the first Russian edition by Leo F. Boron with the assistance of Albert L. Rabenstein and Richard C. Bollinger, P. Noordhoff, Ltd., Groningen, 1964. MR0180763

[65] Aleksey Vasil′yevich Pogorelov, *The Minkowski multidimensional problem*, V. H. Winston & Sons, Washington, D.C.; Halsted Press [John Wiley & Sons], New York-Toronto-London, 1978. Translated from the Russian by Vladimir Oliker; Introduction by Louis Nirenberg; Scripta Series in Mathematics. MR0478079

[66] Rüger, A., *Reflection Coefficients and Azimuthal AVO Analysis in Anisotropic Media*, Geophysical Monograph Series, 2001.

[67] W.V.T. Rusch and P.D. Potter. *Analysis of reflector antennas*, Academic Press, 1970.

[68] W Song, D Cheng, Y Liu and Y. Wang, *Free-form illumination of a refractive surface using multiple-faceted refractors*, Applied Optics, Vol. 54, No. 28, (2015).

[69] Slawinski, M., and Slawinski, R., *Energy partition at the boundary between anisotropic media: Part one* CREWES Research Report , **6**, (1994).

[70] K.P. Thompson and J. Rolland *Freeform Optical Surfaces – A Revolution in Imaging Optical Design*, Optics and Photonics News, 30–37, Invited (2012).

[71] Neil S. Trudinger, *On the local theory of prescribed Jacobian equations*, Discrete Contin. Dyn. Syst. **34** (2014), no. 4, 1663–1681, DOI 10.3934/dcds.2014.34.1663. MR3121636

[72] Neil S. Trudinger, *Recent developments in elliptic partial differential equations of Monge-Ampère type*, International Congress of Mathematicians. Vol. III, Eur. Math. Soc., Zürich, 2006, pp. 291–301. MR2275682

[73] Neil S. Trudinger and Xu-Jia Wang, *On the second boundary value problem for Monge-Ampère type equations and optimal transportation*, Ann. Sc. Norm. Super. Pisa Cl. Sci. (5) **8** (2009), no. 1, 143–174. MR2512204

[74] Cédric Villani, *Topics in optimal transportation*, Graduate Studies in Mathematics, vol. 58, American Mathematical Society, Providence, RI, 2003. MR1964483

[75] Cédric Villani, *Optimal transport. Old and new*, Grundlehren der Mathematischen Wissenschaften [Fundamental Principles of Mathematical Sciences], vol. 338, Springer-Verlag, Berlin, 2009. MR2459454

[76] Xu-Jia Wang, *On the design of a reflector antenna*, Inverse Problems **12** (1996), no. 3, 351–375, DOI 10.1088/0266-5611/12/3/013. MR1391544

[77] Xu-Jia Wang, *On the design of a reflector antenna. II*, Calc. Var. Partial Differential Equations **20** (2004), no. 3, 329–341, DOI 10.1007/s00526-003-0239-4. MR2062947

[78] B. S. Westcott, *Shaped Reflector Antenna Design*, Research Studies Press, Letchworth, UK, 1983.

[79] B. S. Westcott and A. P. Norris , *Reflector synthesis for generalized far-fields*, J. Phys. A: Math. Gen. **8** 521, 1975.

[80] Shing-Tung Yau, *Open problems in geometry*, Differential geometry: partial differential equations on manifolds (Los Angeles, CA, 1990), Proc. Sympos. Pure Math., vol. 54, Amer. Math. Soc., Providence, RI, 1993, pp. 1–28. MR1216573

Department of Mathematics, Howard University, Washington, D.C. 20059
Email address: henok.mawi@howard.edu

Contemporary Mathematics
Volume **759**, 2020
https://doi.org/10.1090/conm/759/15264

Women who count: Using the positive narratives of African American women mathematicians to motivate students and build positive mathematics identities

Shelly M. Jones

Abstract. As an African American woman and a mathematics educator, I was inspired by the movie Hidden Figures to write a book, *Women Who Count: Honoring African American Women Mathematicians* [**13**]. The book aims to bring continued attention to the positive narratives of African American women in mathematics, including their contributions to the field as well as a glimpse into their personal lives. The inspiring biographies of these 29 African American women mathematicians can help to motivate students and support their mathematics identity development. Inspiring role models such as the women profiled in the book and supportive academic environments are two important factors that have helped Black women succeed in mathematics, a STEM field where they continue to be underrepresented. Historically Black Colleges and Universities (HBCUs) have been key in developing successful Black students in STEM by providing peer and faculty role models and supportive environments where students feel a sense of belonging.

1. Introduction

For Black students to be successful at mathematics they need to have positive mathematics identities and they need to see themselves as doers of mathematics. In this paper, I will discuss how students can develop positive mathematics identities through factors such as having inspiring role models, and supportive academic environments [**8**]. I will begin by briefly defining mathematics identity and then provide a vignette revealing how a middle school teacher is helping her students develop positive mathematics identities. I will share data on the success that Historically Black Colleges and Universities (HBCUs) have had in contributing to the number of Black students with STEM degrees myself included. I will end by sharing a sampling of the profiles included in *Women Who Count*, a resource that can be used to expose students to a group of mathematicians that might otherwise be hidden. This resource could help to broaden the representation of women and Blacks in mathematics.

2020 *Mathematics Subject Classification.* Primary 01A70, 01-06, 01A70; Secondary 97-06, 97A40.

2. Mathematics Identity is Crucial

Mathematics identity is defined as the dispositions and deeply held beliefs that students develop about their ability to participate and perform effectively in mathematical contexts and to use mathematics in powerful ways across the contexts of their lives [**1**]. Mathematics identity is a negotiation between how we see ourselves in relation to mathematics and how others position us in relation to mathematics [**22**]. The mathematics classroom; therefore, holds a powerful place in the development of students' mathematics identities because of the contexts teachers choose and the instructional strategies they use [**4, 7, 10, 15, 19, 23**]. Consider the following vignette.

Annie Perkins, a middle school teacher, had this experience when discussing Pythagoras. A student who identifies strongly with his Mexican heritage asked, "Why do we always talk about white dudes?" Unprepared for that question, Perkins asked the student, "Would it matter to you if I showed you a Mexican mathematician?" What happened next was most depressing to her. The student inquired, "Do you think there are any?" The next day, Perkins began what she called the Mathematicians Project, where she introduced a different underrepresented mathematician each week with her class, beginning with mathematician Diego Rodriguez. After the presentation on Diego Rodriguez, the student was so excited that he yelled, "Take that, white dudes!" (`www.arbitrarilyclose.com`)

What Ms. Perkins found was 1) by doing this project her students began to find mathematician role models they could identify with, 2) students wanted to know about the personal lives of the mathematicians, 3) students wanted to learn about mathematicians who very specifically matched their identities (for example, 85% of Black or African students researched Black or African mathematicians), and 4) the teacher learned a lot about her students through this project. In fact, the students created slides of themselves as mathematicians.

In my book, *Women Who Count*, I chose to write biographies about Black women because I was personally attracted to their stories. My story mirrors many of theirs in various ways. One way in particular is the underrepresentation of Black professors in mathematics departments around the country [**11**]. I am the only Black full-time professor in the mathematics department at my university, and I am the first Black faculty member to be promoted to the rank of full professor in the department. Dr. Raegan Higgins, who is featured in the book, has a similar story. She is the first African American to receive tenure and promotion in her university's mathematics and statistics department.

The following data reflect our stories. The National Science Foundation [**17**] reports that Black or African American women continue to be underrepresented in mathematics and other STEM fields. Figure 1 shows that in 2014, Black or African American women earned only 2.35% of bachelor's degrees awarded in mathematics and statistics and that number is down from 3.25% 10 years earlier. Despite these low numbers, HBCUs continue to outpace non-HBCU colleges and universities in successfully graduating Black or African American students in general and in the STEM fields. Mary Schmidt Campbell [**6**], president of Spelman College, reported that Spelman's graduation rate for Black women exceeds the graduation rate for Black women nationally, the graduation rate for Black women at liberal arts colleges in general, and the graduation rate for Black women at women's colleges. Although

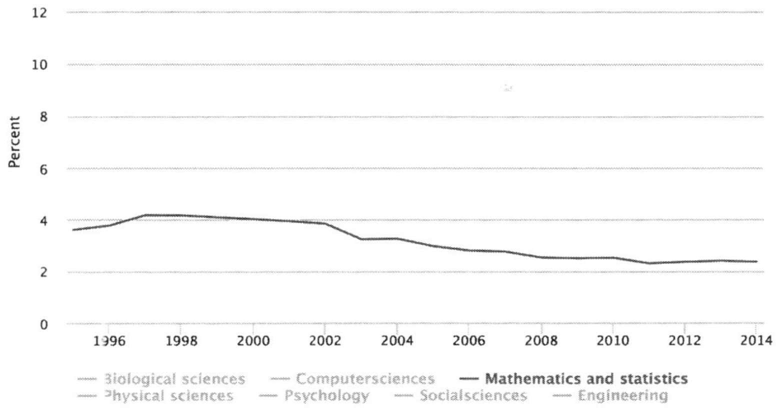

FIGURE 1. Mathematics and Statistics Bachelor's Degrees Earned by Black or African American Women, by Field: 1995–2014 (NSF, 2017).

HBCUs make up just 3% of America's colleges and universities, they produce 25% of African American graduates in the STEM fields [**12**].

3. Historically Black Colleges and Universities: A Key Support for Black Students in STEM

HBCUs provide better learning environments for African American students than do Predominately White Institutions (PWIs), according to a 2015 Gallup study. HBCU graduates report having experienced more support and higher levels of engagement than their Black peers at PWIs. In 2014 HBCUs accounted for 24 percent of the degrees earned by African Americans in STEM fields [**17**]. HBCUs produce large numbers of students with bachelor's degrees who later earn doctoral degrees in mathematics and other science and engineering programs. For example, between 2013 and 2017, 35% of Black graduates who earned a mathematics and statistics doctoral degree earned a bachelor's degree from an HBCU [**18**].

Although these statistics provide evidence of the benefits of an HBCU education, the number of Blacks in mathematics has continued to decline over the past two decades. In Figure 2, we can see that the percentage of Black or African American students (both male and female) earning bachelor's degrees in mathematics and statistics decreased from 7% to 4% between 1996 and 2016 [**18**]. Because mathematics identity is socially constructed, it makes sense that the supportive environment of HBCUs have played such a vital role in the advancement of Black students into the STEM fields.

In [**25**], the authors describe the value of identity construction as a way to strengthen students' mathematics learning especially for students who have been historically underrepresented in disciplines such as science and mathematics: Such students need to be explicitly positioned by themselves and others as capable doers of science and mathematics and in ways that build strong connections between their racial and ethnic identities and their disciplinary identities. Based on my experience and others', this is the work being done at HBCUs.

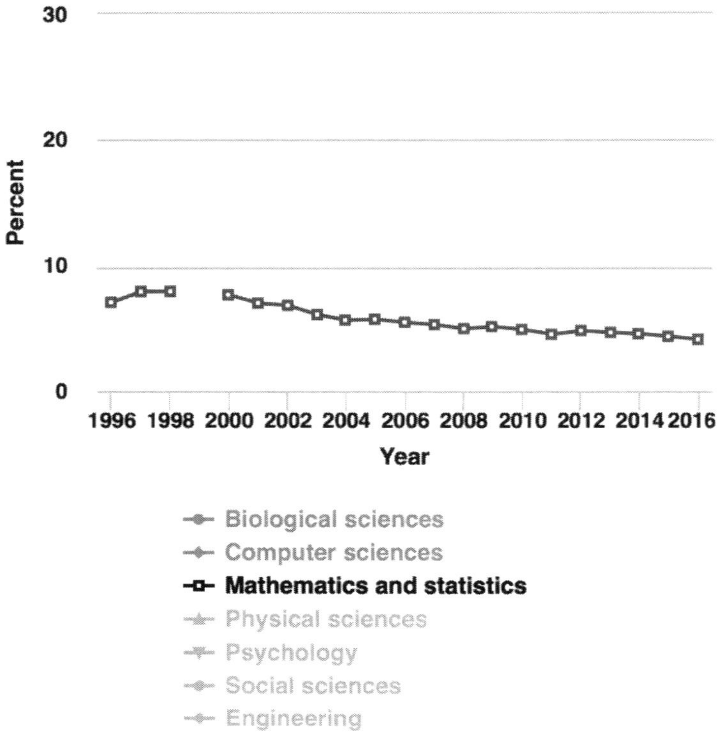

FIGURE 2. Mathematics and Statistics Bachelor's Degrees Earned by Blacks or African Americans as a Percentage of Degree Field. 1996-2016 (National Science Foundation, 2019).

Mathematicians have lauded supportive academic environments like those in HBCUs as being instrumental in their success and perseverance in the STEM field. Another important factor to their success was the presence of an inspiring role model [**5**, **8**]. In [**27**], the author found that Black mathematicians have all types of role models inside and outside of educational settings to whom they attribute their success in the field. Role models might be a family member [**14**], a neighbor [**21**, **27**], a teacher [**9**] or even someone the student never met [**16**].

Dr. Robin Wilson [**26**] described how the website, Mathematicians of the African Diaspora (MAD), motivated him because he felt so isolated as a Black male mathematics student. Until he discovered the site, Dr. Wilson was unaware of the vast number of Black mathematicians. He was frustrated that it was easier to learn about athletes and their achievements than about the intellectual achievements of the Black scholars featured in MAD. In that same vein Sáraco [**20**], a middle school teacher, wanted to explore how to fit minority and women's studies into her mathematics class. She asked her students to name some mathematicians from history. The students had difficulty coming up with five names and of those five, none were female or minority. On the other hand, when asked to name athletes and entertainers, students could name many and knew a great deal about them. I believe that because these students are constantly seeing athletes and entertainers in the media, that is who they know and aspire to be. Many children from underrepresented groups do not see themselves as mathematicians because they are not exposed to mathematicians who look like them. I also believe we can overcome this possible barrier by exposing Black students to profiles of more diverse mathematicians.

Mayes-Tang [**16**] proposes using storytelling as a strategy to introduce female college students to the diverse stories of women mathematicians in an effort to improve the participation of women in mathematics. Others have found cases where middle school [**3**] and high school [**2**], girls were actually turned off by women in STEM because the students felt that they couldn't live up to the high standards of these women. Stories that include personal details about the mathematicians and the struggles they faced in their pursuit of mathematics may help to humanize the mathematicians. and make the idea of becoming a mathematician or other STEM professional less daunting.

4. Women Who Count: Honoring African American Women Mathematicians

I wrote *Women Who Count* in an effort to provide positive narratives of Black women in mathematics, offering a resource to bring Black mathematicians into a variety of settings where all students regardless of gender, race, and ethnicity could learn that brilliance comes in many forms, not just "white dudes." The remainder of this paper offers a sampling of the profiles contained in the book.

Classroom teachers can use the book as a resource to implement culturally relevant mathematics by having students learn about the contributions that diverse people make to mathematics. Gloria Ladson-Billings (2009) defines culturally relevant pedagogy as an approach to teaching that empowers students intellectually, emotionally, socially, and politically. Leonard and her colleagues (2010) argue that culturally relevant instruction can offer opportunities for students to learn mathematics in ways that are deeply meaningful and influential to the development of a positive mathematics identity.

Informal teachers such as those who teach in after school programs, community centers, and academic programs at museums, can incorporate the book into history and mathematics curricula. Parents can motivate their children to succeed in mathematics by reading the book with them, discussing the inspiring stories of the mathematicians, and even engaging in a few of the activities with them. Children can have fun learning about the mathematicians and being challenged by fun math activities. They might also use the book to "play school" or to pass the time away on long road trips. Whatever your role, I hope that the following stories inspire you to continue to learn more about not only African American women mathematicians, but all mathematicians.

5. Counting and More With the Women Who Count

Women Who Count is an activity book geared to children in grades 3 to 8, but it is appropriate for all ages. It includes a portrait sketch (or photograph) and a short biography of each of the 29 featured African American women mathematicians. The book also includes elementary and middle school level mathematics activities such as word searches of mathematicians' names, crossword puzzles of mathematics facts, scrambled mathematics vocabulary words, magic squares and equations to solve, coloring pages, an eye-spy science lab, and more.

The book is divided into four sections of mathematicians. The first section, "The Firsts", includes the first three African American women to earn doctorates in mathematics: Drs. Martha Euphemia Lofton Haynes, Evelyn Boyd Granville, and Marjorie Lee Browne. The second section, "The Pioneers", includes nine African American women mathematicians who laid the groundwork and became role models for future mathematicians. The third section, "The UnHidden Figures", includes four distinguished women whose lives were chronicled in the book *Hidden Figures: The American Dream and the Untold Story of the Black Women Mathematicians Who Helped Win the Space Race* by Margot Lee Shetterly which later became the basis for a major motion picture. The final section, "The Contemporary Firsts", includes 13 modern mathematicians who are breaking barriers by becoming firsts in their universities and companies.

5.1. The Firsts. The first African American woman to earn a doctorate in mathematics was Martha Euphemia Lofton Haynes. Euphemia, as she is best known, grew up in Washington, D.C. She earned a bachelor's degree in mathematics from the all-female Smith College in 1914, and a master's degree in education from the University of Chicago in 1930. After receiving her master's degree, she joined the faculty of Miner Teacher's College, which stressed training African American teachers. She went on to earn her doctoral degree from Catholic University in 1943, and spent nearly 50 years as an educator. Dr. Haynes died in 1980.

Evelyn Boyd Granville was the second African American woman to earn a doctorate in mathematics, and she received the first honorary doctorate given by an American college to an African American woman in mathematics, from Smith College. A mathematician, computer scientist, and educator, Dr. Granville was also from Washington, D.C., and still lives in the area. Like Dr. Haynes, she earned a bachelor's degree in mathematics from Smith College in 1945. She then completed double master's degrees in physics and mathematics in 1946 at Yale University, and earned the doctorate at Yale in 1949. In addition to the groundbreaking honorary degree from Smith, she has honorary doctorates from Spelman College, Yale University, and Lincoln University. In January of 2019, at the age of 94, Dr. Boyd attended the Annual National Association of Mathematicians Banquet and was awarded the NAM Golden Anniversary Legacy Award.

Marjorie Lee Browne was the third African American woman to earn a doctorate in mathematics. She earned a bachelor's degree in mathematics from Howard University in 1935, a master's degree from the University of Michigan in 1939, and a doctoral degree in mathematics from the University of Michigan in 1950. After receiving her doctorate, Dr. Browne taught at North Carolina Central University, the nation's first public liberal arts college founded for African Americans. She died in 1979 at the young age of 65. In 2001 the University of Michigan began a lecture series honoring Dr. Browne.

5.2. The Pioneers. The pioneer section of the book includes nine women who picked up the torch lit by The Firsts. These women were all role models for the next generation, giving of themselves to broaden the participation of minorities and women in mathematics. Dr. Etta Zuber Falconer, for example, positively impacted the lives of hundreds of students in the STEM fields during her 37-year career at Spelman College. In 1953 Dr. Falconer graduated with a bachelor's degree in mathematics from Fisk University, earned a master's degree in mathematics from the University of Wisconsin–Madison in 1954, and in 1969 became the tenth African American woman to earn a doctorate in mathematics, from Emory University. For her dedication and mentorship Dr. Falconer received many awards and honors. In 2002 Dr. Falconer retired from Spelman. She died in 2002 at the young age of 68.

Dr. Sylvia T. Bozeman co-founded a national program, Enhancing Diversity in Graduate Education (EDGE), with her colleague Rhonda Hughes of Bryn Mawr College in 1998. The goal of EDGE is to assist women in making the transition to graduate school and in earning doctoral degrees in the mathematical sciences. Dr. Bozeman earned three degrees in mathematics: a bachelor's degree from the HBCU Alabama A&M University in 1968, a master's degree from Vanderbilt University in 1971, and a doctoral degree from Emory University in 1980. She worked most of her career at Spelman College, where she is retired. Dr. Bozeman received the first Mathematical Association of America Diversity and Inclusion Award in 2019.

Dr. Genevieve Knight teamed with six other mathematics educators in 1986 to found the Benjamin Banneker Association, Incorporated (BBA), which is dedicated to providing the highest quality mathematics education for Black students. Dr. Knight earned a bachelor's degree in mathematics from Fort Valley State College in 1961, a master's degree in mathematics from Atlanta University in 1963, and a doctorate in mathematics education from the University of Maryland in 1970. Dedicated to teaching at HBCUs, she taught at Hampton Institute (now Hampton University), and at Coppin State College from 1985 until she retired in 2006. For her 36 years of service to mathematics education, the National Council of Teachers of Mathematics honored her with its Lifetime Achievement Award in 1999.

5.3. The UnHidden Figures. The UnHidden Figures includes four distinguished women whose lives were chronicled in the book Hidden Figures by Margot Lee Shetterly. These women, among many others, played significant roles in the space race, as "human computers" at the National Aeronautical Space Administration (NASA). One of those women, Dr. Christine Darden (Darden, 2019), said that she likes that all the women of Hidden Figures stood up for themselves. Katherine Johnson fought to attend the meetings with NASA engineers who were all white men, Mary Jackson won the right to take higher-level mathematics classes at an all-white segregated high school, and Dorothy Vaughan learned how to program the IBM computer on her own when she found out the computer might replace her job. She also trained her staff to program the computers.

Dr. Darden said she stood up for herself when her father wanted her to get a teaching certificate but she wanted to become an engineer. She listened to her father and enrolled in college to become a teacher; however, in addition to her education courses, she took four higher-level mathematics classes during each of her last two semesters in order to also complete an undergraduate mathematics degree. It is important to note that one of the semesters was during her student teaching, and the three hours of classes were at a university 80 miles away. Students need to hear about the struggles of those who came before them to know that these women of color fought hard, persevered, and succeeded. After she earned that bachelor's degree in mathematics and teaching certificate from Hampton Institute (now Hampton University), she taught high school mathematics. She earned a master's degree in applied mathematics from Virginia State College (now Virginia

State University) in 1967, and that same year was hired at NASA as a data analyst. While at NASA, she fought to become an engineer because she found that her applied mathematics was very close to what engineers did, and she saw men with the same credentials she had get promoted before her. In 1973 she was promoted to the position of aerospace engineer. Dr. Darden continued her education at George Washington University, and in 1983 earned a doctorate in mechanical engineering. During her 40-year career at NASA she was a leading researcher in sonic boom minimization.

Katherine Johnson, another of the women profiled in this section, graduated from high school at the young age of 14 and in 1937 earned two bachelor's degrees from West Virginia State College (now West Virginia State University), in French and mathematics, by the age of only 18. She began her professional career as a high school mathematics teacher but is most known for her work at NASA. In 1953 she joined NACA's Langley Memorial Aeronautical Laboratory as a research mathematician. She started in the segregated West Computing section, where she calculated the flight trajectory for Alan Shepard, the first American to go into space in 1959. She verified the mathematics behind astronaut John Glenn's orbit around the Earth in 1962, and is known for calculating the flight trajectory for Apollo 11's flight to the moon in 1969. Mrs. Johnson retired from NASA in 1986, and has received many honors and awards for her important work there. Mrs. Johnson dies in 2020 at the age of 101.

5.4. The Contemporary Firsts. In the Contemporary Firsts section, I have featured colleagues including Drs. Raegan Higgins, Monica Jackson, Yolanda Parker, Candice Price, Erica Walker, Chelsea Walton, Shelby Wilson, and Talitha Washington. When students learn about these dynamic women, they will be inspired by someone who seems reachable. These women can be found on social media and at conferences in your town.

Dr. Talitha Washington, for example, is a mathematician, professor, activist, and sought-after motivational speaker because of her commitment to diversity in the STEM field. Dr. Washington is the inaugural Director of the Data Science Initiative at the Atlanta University Center Consortium, Inc. She was formerly an associate professor of mathematics at Howard University, and a program officer in the Division of Undergraduate Education at the National Science Foundation. She earned an undergraduate mathematics degree at Spelman College and master's and doctoral degrees from the University of Connecticut, where in 2001 she became their first African American woman to earn a mathematics doctorate [**28**]. Dr. Washington has received many awards, most recently the 2019 Black Engineer of the Year STEM Innovator Award and the Outstanding Faculty Award (2018-2019) from the College of Arts and Sciences at Howard University. With her passion for education she led a youth conference, Stepping Up, that encouraged youth to pursue viable careers through higher education. She also led a one-week, research-based summer camp for middle schoolers to explore current trends in mathematics and the sciences. She has given several talks on the mathematics of the women in the book, *Hidden Figures*.

Dr. Raegan Higgins has experienced some important firsts in her life. In 2008 she was one of the first two African American women to earn a doctoral degree in mathematics from the University of Nebraska–Lincoln. Currently she is an associate professor of mathematics at Texas Tech University, where she was the first African American to receive tenure and promotion in the mathematics and statistics department. An innovator, Dr. Higgins is one of four creators of the website Mathematically Gifted and Black, whose mission is to feature and share the accomplishments of Blacks in the mathematical sciences. She is also a co-director of the EDGE Program. Dr. Higgins believes it is important to show women, especially African American and Hispanic women, that they can succeed in the mathematical sciences.

6. Excellence Inspires Excellence

Many of the mathematicians featured in Women Who Count are graduates of HBCUs (see Table 1). Prior to the 1960s, HBCUs were some of the only institutions of higher education available to Blacks, so a majority of The Pioneers attended HBCUs. Today HBCUs enroll 11% of Black students in the United States, even though they represent less than 3% of colleges and universities in the country [**24**]. My personal experience at an HBCU was an experience that cannot be overstated. I was led to believe that I could do anything that I worked hard to achieve. Being surrounded by Black excellence in the faculty, students, and staff propelled me to excel as well. Whether or not the women in the book attended an HBCU did not determine their success; however, a common theme for all the women was that they all experienced a place that had a sense of belonging in their educational and/or professional careers. For example, Erica Graham told participants at the 2017 Annual Meeting of the Society for Industrial and Applied Mathematics (SIAM) that when she was a graduate student, the EDGE program afforded her the opportunity to meet Black women in mathematics who had already achieved what she was getting ready to do, pursue a mathematics graduate degree. During my career, I have been blessed to belong to the Benjamin Banneker Association, a professional organization that has allowed me to establish relationships with other Black colleagues who have the same passion I have, to improve the success of Black students in STEM, and to thereby increase our representation in the field.

7. Final Comments

The mathematicians in Women Who Count and many other women have made and continue to make strides in a variety of academic, industrial, and government careers. It is important to engage children at all grade levels with a more diverse representation of people in STEM, including women and people of color. This paper provides parents and educators a variety of resources to assist them with introducing children to diverse people in STEM, beginning with brief stories about African American women mathematicians. Parents and educators can further explore and broaden their options by reading the book, Women Who Count, and by visiting these websites: Mathematically Gifted and Black (`http://mathematicallygiftedandblack.com/`) and Mathematicians of the African Diaspora (`http://www.math.buffalo.edu/mad/`)

In addition to providing students with role models, it is also important to provide supportive environments where students' mathematics identities can be developed. HBCUs have been recognized as one place where this is happening. How can we replicate the supportive environment of the HBCU in other educational settings? This question is beyond the scope of this paper; however, it is one worth considering. Organizations such as The Benjamin Banneker Association

TABLE 1. Undergraduate Education of Mathematicians Featured in Women Who Count

Undergraduate Institution (HBCU*)	**Mathematicians**
Alabama A&M University*	Sylvia Bozeman
Birmingham-Southern College	Erica Walker
California State University	Candice Price
Clark Atlanta University*	Monica Jackson, Shree Taylor
Fisk University*	Etta Z. Falconer, Gloria Hewitt, Vivienne Mayes
Fort Valley State College* (now University)	Genevieve Knight
Grambling State University*	Talea Mayo
Hampton Institute* (now University)	Christine Darden, Mary Jackson
Howard University*	Marjorie Browne
LeMoyne College*	Sadie Gasaway
Marianao Institute, Havana Cuba	Argelia Velez Rodriguez
Michigan State University	Chelsea Walton
Morgan State University*	Gloria Gilmer
Smith College	Evelyn Granville, Euphemia Haynes
Spelman College*	Talitha Washington, Kimberly Weems, Shelby Wilson
Texas A&M University	Yolanda Parker
West Chester University	Carol Malloy
West Virginia State College* (now University)	Katherine Johnson
Wilberforce University*	Dorothy Vaughan
Xavier University*	Christina Eubanks-Turner, Raegan Higgins, Tasha Inniss

and The National Association of Mathematicians have a track record of supporting Black mathematicians. The Enhancing Diversity in Graduate Education Program has also been recognized as offering an environment that allows its participants, women of color, to thrive. *Sankofa* is an African word that is translated as, "it is not taboo to go back and fetch what you forgot." The idea of Sankofa teaches us that we must go back to our roots in order to move forward. My reaction to the movie Hidden Figures lead to a Sankofa moment. I was exasperated and perplexed that I hadn't learned about the important contributions of the Black women portrayed in the movie. From that moment I was inspired to do my own search of other hidden figures. My search resulted in Women Who Count. The stories will introduce students to the contributions Black women have made to the STEM fields. Reading the stories in the first three sections of book, The Firsts, The Pioneers, and The Unhidden Figures, takes students back in history to learn how these women succeeded even when faced with obstacles. Reading about the women in The Contemporary Firsts section could be especially influential, because the personal information these mathematicians share in their stories make them more identifiable to young people. Students must see themselves in mathematics and must feel a sense of belonging to build the positive mathematics identities that result in their persistence in STEM thereby increasing and enriching diversity in the STEM fields.

References

[1] J. Aguirre, K. Mayfield-Ingram, and D. B. Martin, *The impact of identity in K-8 mathematics learning and teaching: Rethinking equity-based practices*, Reston, VA, 2013.

[2] Y. M. Bamberger, *Encouraging girls into science and technology with feminine role model: Does this work?*, Journal of Science Education and Technology **23** (2014), no. 4, 549–561.

[3] D. B. Betz and D. Sekaquaptewa, *My fair physicist? feminine math and science role models demotivate young girls*, Social Psychological and Personality Science **3** (2012), no. 6, 738–746.

[4] J. Boaler and M. Staples, *Creating mathematical futures through an equitable teaching approach: The case of railside school*, Teachers College Record **110** (2008), no. 3, 608–645.

[5] V. Borum and E. N. Walker, *What makes the difference? Black women's undergraduate and graduate experiences in mathematics*, The Journal of Negro Education **81** (2012), no. 4, 366–378.

[6] M. S. Campbell, *Spelman president responds to the inaccurate portrayal of HBCUs by the Atlanta Journal-Constitution*, The Atlanta Journal Constitution (2018February).

[7] T. Chao and D Jones, *That's not fair and why: Developing social justice activists in prek*, Mathematics Teaching in the Middle School **14** (2016), no. 2, 70–76.

[8] K. S. Cohen, *Unhidden figures*, SIAM News **50** (2017), no. 8, 8–9.

[9] S. Gershenson, C. M. D. Hart, C. A. Lindsay, and N. W. Papageorge, *The long-run impacts of same-race teachers*, IZA Institute of Labor Economics Discussion Paper Series, Bonn, Germany, 2019.

[10] G. Gutstein, P. Lipman, P. Hernandez, and T. de los Reyes, *Culturally relevant mathematics teaching in a Mexican American contextl*, Journal for Research in Mathematics Education **28** (1997), no. 6, 709–737.

[11] A Harmon, *For a black mathematician, what it's like to be the 'only one'*, New York Times (2019February).

[12] J. Humphreys, *Hbcus make america strong: The positive economic impact of historically black colleges and universities*, Washington, DC, 2017.

[13] S. M. Jones, *Women who count*, American Mathematical Society, Providence, RI, 2019. Honoring African American women mathematicians; Illustrated by Veronico Martins. MR3966443

[14] P. C. Kenschaft, *Black women in mathematics in the United States*, Amer. Math. Monthly **88** (1981), no. 8 592–604, DOI 10.2307/2320508. MR628027

[15] L. E. Matthews, S. M. Jones, and Y. A. Parker, *Advancing a framework for culturally relevant, cognitively demanding mathematics tasks*, The brilliance of Black children in mathematics: Beyond the numbers and toward a new discourse, 2013, pp. 123–150.

[16] S. Mayes-Tang, *Telling women's stories: A resource for college mathematics instructors*, Journal of Humanistic Mathematics **9** (2019), no. 2, 78–92.

[17] National Science Foundation, *National center for science and engineering statistics, women, minorities, and persons with disabilities in science and engineering*, Arlington, VA, 2017.

[18] National Science Foundation, *National center for science and engineering statistics, women, minorities, and persons with disabilities in science and engineering*, Arlington, VA, 2019.

[19] L. H. Rubel, *Equity-directed instructional practices: Beyond the dominant perspective*, Journal of Urban Mathematics Education **10** (2017), no. 2, 66–105.

[20] M. R. Sáraco, *Historical research: How to fit minority and women's studies into mathematics class*, Mathematics Teaching in the Middle School **14** (2008), no. 2, 70–76.

[21] M. L. Shetterly, *Hidden figures: The american dream and the untold story of the black women mathematicians who helped win the space race*, William Morrow, an Imprint of Harper Collins, New York, NY, 2016.

[22] J. Spencer, *The solitude of prime numbers [book review, Pamela Dorman Books, New York, 2010]*, Notices Amer. Math. Soc. **57** (2010), no. 8, 980–981. MR2667496

[23] E. E. Turner and B. T. Font Strawhun, *Posing problems that matter: Investigating school overcrowding*, Teaching Children Mathematics **13** (2007), no. 9, 457–463.

[24] U.S. Department of Education, *Integrated postsecondary education data system hbcu faculty and salary component*, Washington, DC, 2011.

[25] M. Varelas, D. B. Martin, and J. M. Kane, *Content learning and identity construction: A framework to strengthen african american students' mathematics and science learning in urban elementary schools*, Human Development **55** (2012), no. 5, 319–339.

[26] E. Walker, S. Williams, and R. Wilson, *An existence proof: The mathematicians of the African diaspora website– part I*, National Association of Mathematicians Newsletter **4** (2018), 10–12.

[27] E. N. Walker, *Beyond Banneker: Black mathematicians and the paths to excellence*, State University of New York Press, Albany, NYI, 2014.

[28] T. M. Washington, *Behind every successful woman, there are a few good men*, Notices Amer. Math. Soc. **65** (2018), no. 2, 132–134, DOI 10.1090/noti1634. MR3751308

Department of Mathematical Sciences, Central Connecticut State University, New Britian, CT 06050

Email address: jonessem@ccsu.edu

Contemporary Mathematics
Volume **759**, 2020
https://doi.org/10.1090/conm/759/15265

A constructive proof of Masser's Theorem

Alexander J. Barrios

ABSTRACT. The Modified Szpiro Conjecture, equivalent to the *abc* Conjecture, states that for each $\epsilon > 0$, there are finitely many rational elliptic curves satisfying $N_E^{6+\epsilon} < \max\{|c_4^3|, c_6^2\}$ where c_4 and c_6 are the invariants associated to a minimal model of E and N_E is the conductor of E. We say E is a good elliptic curve if $N_E^6 < \max\{|c_4^3|, c_6^2\}$. Masser showed that there are infinitely many good Frey curves. Here we give a constructive proof of this assertion.

1. Introduction

By an ABC triple, we mean a triple of positive integers (a, b, c) such that a, b, and c are relatively prime positive integers with $a + b = c$. The ABC Conjecture [**CR01**, 5.1] states that for any $\epsilon > 0$, there are only finitely many ABC triples that satisfy $\operatorname{rad}(abc)^{1+\epsilon} < c$ where $\operatorname{rad}(n)$ denotes the product of the distinct primes dividing n. We say that an ABC triple is good if $\operatorname{rad}(abc) < c$. For instance, the triple $(1, 8, 9)$ is a good ABC triple and more generally the triple $(1, 9^k - 1, 9^k)$ is a good ABC triple for each positive integer k [**CR01**]. In 1988, Oesterlé [**Oes88**] proved that the ABC Conjecture is equivalent to the modified Szpiro conjecture which states that for $\epsilon > 0$, there are only finitely many elliptic curves E such that $N_E^{6+\epsilon} < \max\{|c_4^3|, c_6^2\}$ where N_E denotes the conductor of the elliptic curve and c_4 and c_6 are the invariants associated to a minimal model of E. As with ABC triples, we define a good elliptic curve to be an elliptic curve E that satisfies the inequality $N_E^6 < \max\{|c_4^3|, c_6^2\}$. In the special case of Frey curves, that is, a rational elliptic curve that has a Weierstrass model of the form $y^2 = x(x-a)(x+b)$ where a and b are relatively prime integers, Masser [**Mas90**] showed that there are infinitely many good Frey curves. In this article, we provide a constructive proof of Masser's Theorem. Moreover, the torsion subgroup of a Frey curve can only take on four possibilities due to Mazur's Torsion Theorem [**Maz77**], namely $E(\mathbb{Q})_{\text{tors}} \cong C_2 \times C_{2N}$ where C_m denotes the cyclic group of order m and $N = 1, 2, 3$, or 4. With this we state our main theorem:

Theorem 1 For each of the four possible torsion subgroups $T = C_2 \times C_{2N}$ where $N = 1, 2, 3$, or 4, there are infinitely many good elliptic curves such that $E(\mathbb{Q})_{\text{tors}} \cong T$.

This is equivalent to Theorem 6.3, where the main theorem is given in its constructive form. As a consequence we get examples akin to the infinitely many

2020 *Mathematics Subject Classification.* Primary 11G05.
Key words and phrases. Number theory, elliptic curves, arithmetic geometry.

good ABC triples $(1, 9^k - 1, 9^k)$ for each positive integer k. For each of the four possible T, we use rational maps of modular curves to construct a recursive sequence of ABC triples $P_j^T = (a_j, b_j, c_j)$ such that if P_j^T is a good ABC triple satisfying certain congruences, then P_j^T is a good ABC triple for each nonnegative integer j. Once this is proven, we prove our main Theorem by showing that the associated Frey curve

$$F_{P_j^T} : y^2 = x(x - a_j)(x + b_j)$$

is a good elliptic curve for each positive integer j with $F_{P_j^T}(\mathbb{Q})_{\text{tors}} \cong T$.

2. Certain polynomials

In this section we establish a series of technical results which will ease the proofs in the sections that are to follow. Let $T = C_2 \times C_{2N}$ where $N = 1, 2, 3, 4$. For each T let $\mathfrak{A}_T = \mathfrak{A}_T(a, b)$, $\mathfrak{B}_T = \mathfrak{B}_T(a, b)$, $\mathfrak{C}_T = \mathfrak{C}_T(a, b)$, $\mathfrak{D}_T = \mathfrak{D}_T(a, b)$, $\mathfrak{A}_T^r = \mathfrak{A}_T^r(a, b)$, $\mathfrak{B}_T^r = \mathfrak{B}_T^r(a, b)$, $\mathfrak{C}_T^r = \mathfrak{C}_T^r(a, b)$, $U_T = U_T(a, b, r, s)$, $V_T = V_T(a, b, r, s)$, and $W_T = W_T(a, b, r, s)$ be the polynomials in $R = \mathbb{Z}[a, b, r, s]$ defined in Table 6.

For a fixed T, the polynomials $\mathfrak{A}_T$, $\mathfrak{B}_T$, $\mathfrak{C}_T$, and $\mathfrak{D}_T$ are homogenous polynomials in a and b of the same degree m_T. In particular, we have the equalities

$$a^{m_T}\mathfrak{A}_T\left(1, \tfrac{b}{a}\right) = \mathfrak{A}_T(a, b) \qquad a^{m_T}\mathfrak{B}_T\left(1, \tfrac{b}{a}\right) = \mathfrak{B}_T(a, b)$$
$$a^{m_T}\mathfrak{C}_T\left(1, \tfrac{b}{a}\right) = \mathfrak{C}_T(a, b) \qquad a^{m_T}\mathfrak{D}_T\left(1, \tfrac{b}{a}\right) = \mathfrak{D}_T(a, b).$$

The first result can be verified via a computer algebra system and we note that we are considering $\mathfrak{A}_T(1, t)$, $\mathfrak{B}_T(1, t)$, $\mathfrak{C}_T(1, t)$, $\mathfrak{D}_T(1, t)$ as functions from $\mathbb{R}$ to $\mathbb{R}$.

LEMMA 2.1. For $T = C_2 \times C_{2N}$ with $N = 1, 2, 3, 4$, let $f_T, g_T : \mathbb{R} \to \mathbb{R}$ be the function in the variable t defined in Table 6. Let θ_T be the greatest real root of $f_T(t)$. The (approximate) value of θ_T is found in Table 6. Then for each T,

(1) $\mathfrak{A}_T \in 4R$;
(2) $\mathfrak{A}_T + \mathfrak{B}_T = \mathfrak{C}_T$;
(3) $U_T\mathfrak{B}_T + V_T\mathfrak{C}_T = W_T$;
(4) $f_T\left(\frac{b}{a}\right) = \frac{\mathfrak{B}_T(a,b)}{\mathfrak{A}_T(a,b)} - \frac{b}{a}$;
(5) $g_T(t) = \mathfrak{C}_T(1, t) - \mathfrak{D}_T(1, t)$;
(6) $f_T(t), g_T(t)$, $\mathfrak{A}_T(1, t)$, $\mathfrak{B}_T(1, t)$, $\mathfrak{C}_T(1, t)$, $\mathfrak{D}_T(1, t) > 0$ for $t > \theta_T$;
(7) For $T = C_2 \times C_{2N}$ for $N = 1, 2$, $f_T(t), g_T(t)$, $\mathfrak{A}_T(1, t)$, $\mathfrak{B}_T(1, t)$, $\mathfrak{C}_T(1, t)$, $\mathfrak{D}_T(1, t) > 0$ for t in $(0, 1)$.

3. Good ABC triples

DEFINITION 3.1. By an ABC triple, we mean a triple $P = (a, b, c)$ such that a, b, and c are relatively prime positive integers with $a + b = c$. We say $P = (a, b, c)$ is good if $\operatorname{rad}(abc) < c$.

LEMMA 3.2. For each $T = C_2 \times C_{2N}$, let $P = (a, b, a + b)$ be an ABC triple with a even and $\frac{b}{a} > \theta_T$ where θ_T is as defined in Lemma 2.1. Suppose further that $a \equiv 0 \mod 3$ if $N = 3$. Then $(\mathfrak{A}_T, \mathfrak{B}_T, \mathfrak{C}_T)$ is an ABC triple with $\mathfrak{A}_T \equiv 0 \mod 16$, $\mathfrak{B}_T \equiv 1 \mod 4$, and $\frac{\mathfrak{B}_T}{\mathfrak{A}_T} > \theta_T$. Moreover, if $N = 3$, then $\mathfrak{A}_T \equiv 0 \mod 3$.

PROOF. Since a and b are relatively prime, there exist integers r and s such that $ra^n + sb^n = 1$, for any positive integer n. Therefore, by Lemma 2.1, $\gcd(\mathfrak{B}_T, \mathfrak{C}_T)$

divides 32 if $N \neq 3$ and $\gcd(\mathfrak{B}_T, \mathfrak{C}_T)$ divides 48 if $N = 3$. Since a is even and $a \equiv 0 \mod 3$ when $N = 3$, we conclude that $\gcd(\mathfrak{B}_T, \mathfrak{C}_T) = 1$. Next, observe that

$$f_T\left(\frac{b}{a}\right) = \frac{\mathfrak{B}_T(1, \frac{b}{a})}{\mathfrak{A}_T(1, \frac{b}{a})} - \frac{b}{a}.$$

Since $\frac{b}{a} > \theta_T$, we have by Lemma 2.1 that $f_T(\frac{b}{a})$ is positive and therefore $\frac{\mathfrak{B}_T}{\mathfrak{A}_T} > \frac{b}{a} > \theta_T$. By Lemma 2.1 we also have that $\mathfrak{A}_T + \mathfrak{B}_T = \mathfrak{C}_T$ for each T and therefore $(\mathfrak{A}_T, \mathfrak{B}_T, \mathfrak{C}_T)$ is an ABC triple. Since a is even it is easily verified that $\mathfrak{A}_T \equiv 0 \mod 16$. Similarly, when $N = 3$, $\mathfrak{A}_T \equiv 0 \mod 3$ since $a \equiv 0 \mod 3$. It easily checked that for each T, $\mathfrak{B}_T \equiv b^{2k} \mod 4$ for some integer k. Since b is odd, it follows that $\mathfrak{B}_T \equiv 1 \mod 4$. □

LEMMA 3.3. Let $P = (a, b, a + b)$ be a good ABC triple and assume the statement of Lemma 3.2. Then $(\mathfrak{A}_T, \mathfrak{B}_T, \mathfrak{C}_T)$ is a good ABC triple.

PROOF. Since a is assumed to be even, we have that $\mathrm{rad}(2^n ax) = \mathrm{rad}(ax)$ for some integer x. Therefore

$$\mathrm{rad}(\mathfrak{A}_T) = \mathrm{rad}(\mathfrak{A}_T^r), \qquad \mathrm{rad}(\mathfrak{B}_T) = \mathrm{rad}(\mathfrak{B}_T^r), \qquad \mathrm{rad}(\mathfrak{C}_T) = \mathrm{rad}(\mathfrak{C}_T^r).$$

Since $(a, b, a + b)$ is a good ABC triple, we have that $\mathrm{rad}(ab(a + b)) < a + b$. From this and the fact that $\mathrm{rad}(xy^k) = \mathrm{rad}(xy) \leq xy$ for positive integers k, x, y, we have that for each T, we attain

$$\mathrm{rad}(\mathfrak{A}_T\mathfrak{B}_T\mathfrak{C}_T) = \mathrm{rad}(\mathfrak{A}_T^r\mathfrak{B}_T^r\mathfrak{C}_T^r) < |\mathfrak{D}_T|.$$

Since $\frac{b}{a} > \theta_T$, $\mathfrak{D}_T(1, \frac{b}{a})$ is positive by Lemma 2.1. In particular, $\mathfrak{D}_T$ is positive since $a^{m_T}\mathfrak{D}_T(1, \frac{b}{a}) = \mathfrak{D}_T$ where m_T is the homogenous degree of $\mathfrak{D}_T$. Now observe that

$$\mathfrak{C}_T - \mathrm{rad}(\mathfrak{A}_T\mathfrak{B}_T\mathfrak{C}_T) > \mathfrak{C}_T - \mathfrak{D}_T = a^{m_T}\left(\mathfrak{C}_T\left(1, \frac{b}{a}\right) - \mathfrak{D}_T\left(1, \frac{b}{a}\right)\right) > 0$$

where the positivity follows from Lemma 2.1. Hence $(\mathfrak{A}_T, \mathfrak{B}_T, \mathfrak{C}_T)$ is a good ABC triple since $\mathrm{rad}(\mathfrak{A}_T\mathfrak{B}_T\mathfrak{C}_T) < \mathfrak{C}_T$. □

PROPOSITION 3.4. Let (a_0, b_0, c_0) be a good ABC triple with a_0 even. For each T define the triple P_j^T recursively by

$$P_j^T = (a_j, b_j, c_j) = (\mathfrak{A}_T(a_{j-1}, b_{j-1}), \mathfrak{B}_T(a_{j-1}, b_{j-1}), \mathfrak{C}_T(a_{j-1}, b_{j-1})) \qquad \text{for } j \geq 1.$$

Assume further that $\frac{b_0}{a_0} > \theta_T$ and that $b_0 \equiv 0 \mod 3$ if $T = C_2 \times C_6$. Then for each $j \geq 1$, P_j^T is a good ABC triple with $a_j \equiv 0 \mod 16$, $b_j \equiv 1 \mod 4$, and $\frac{b_j}{a_j} > \theta_T$. Additionally, if $T = C_2 \times C_6$, then $a_j \equiv 0 \mod 3$.

PROOF. This follows automatically from Lemmas 3.2 and 3.3. □

4. Frey curves

As before, we suppose $T = C_2 \times C_{2N}$ and define for $t \in \mathbb{P}^1$, the mapping $\mathcal{X}_t$ as the mapping which takes T to the elliptic curve $\mathcal{X}_t(T)$ where the Weierstrass model of $\mathcal{X}_t(T)$ is given in Table 1. Our parameterizations for $T = C_2 \times C_{2N}$ where $N = 3, 4$ are those found in [**HLP00**, Table 3] which expands the implicit expressions for the parameters b and c in [**Kub76**, Table 3] to express the universal elliptic curves for the modular curves $X_1(2, 2N)$ in terms of a single parameter t.

Similarly, our model for $T = C_2 \times C_4$ differs by a linear change of variables from the model given for W_4 in [**Sil97**, §4] which parameterizes elliptic curves E with $C_4 \times C_4 \hookrightarrow E(\mathbb{Q}(i))_{\text{tors}}$. In particular, $\mathcal{X}_t(T)$ is a one-parameter family of elliptic curves with the property that if $t \in K$ for some field K, then $\mathcal{X}_t(T)$ is an elliptic curve over K and $T \hookrightarrow \mathcal{X}_t(T)(K)_{\text{tors}}$.

TABLE 1. Universal Elliptic Curve $\mathcal{X}_t(T)$

$\mathcal{X}_t(T) : y^2 + (1-g)\,xy - fy = x^3 - fx^2$		
f	g	T
$\frac{2t^4-7t^3+12t^2-7t+2}{2(-1+t)^4}$	1	$C_2 \times C_4$
$\frac{-2t^3+14t^2-22t+10}{(t+3)^2(t-3)^2}$	$\frac{-2t+10}{(t+3)(t-3)}$	$C_2 \times C_6$
$\frac{16t^3+16t^2+6t+1}{(8t^2-1)^2}$	$\frac{16t^3+16t^2+6t+1}{2t(4t+1)(8t^2-1)}$	$C_2 \times C_8$

For $T = C_2 \times C_2$, define

$$\mathcal{X}_t(T) : y^2 = x^3 + \left(t^4 - 12t^3 + 6t^2 - 12t + 1\right)x^2 - 8t\,(t-1)^4\left(t^2+1\right)x. \tag{4.1}$$

LEMMA 4.1. If $t \in \mathbb{Q}$ such that $\mathcal{X}_t(T)$ is an elliptic curve, then $T \hookrightarrow \mathcal{X}_t(T)(\mathbb{Q})_{\text{tors}}$.

PROOF. Recall that the modular curve $X_1(2,2N)$ (with cusps removed) for $N = 2,3,4$ parameterizes isomorphism classes of pairs (E,P,Q) where E is an elliptic curve having full 2-torsion, P and Q are torsion points of order 2 and $2N$, respectively, and $\langle P, N \cdot Q\rangle = E[2]$.

For $T = C_2 \times C_{2N}$ where $N = 3,4$, we note that our parameterizations are those of the universal elliptic curve for the modular curve $X_1(2,2N)$ [**HLP00**, Table 3]. Thus $T \hookrightarrow \mathcal{X}_t(T)\,(\mathbb{Q})_{\text{tors}}$.

For $T = C_2 \times C_4$, let

$$t' = \frac{t}{2\,(t-1)^2}$$

so that $\mathcal{X}_t(T)$ is equal to the Weierstrass model given for the universal elliptic curve over $X_1(2,4)$ given in [**HLP00**, Table 3] with parameter t'. Hence $T \hookrightarrow \mathcal{X}_t(T)(\mathbb{Q})_{\text{tors}}$.

For $T = C_2 \times C_2$, let $t = \frac{b}{a}$ and consider the admissible change of variables $x \longmapsto \frac{1}{a^4}x$ and $y \longmapsto \frac{1}{a^6}y$. This gives a $\mathbb{Q}$-isomorphism between $\mathcal{X}_t(T)$ and the elliptic curve

$$y^2 = x\left(x - 8ab\left(a^2+b^2\right)\right)\left(x + (a-b)^4\right)$$

which has $\left\langle \left(8ab\left(a^2+b^2\right),0\right),(0,0)\right\rangle \cong C_2 \times C_2$. Thus $T \hookrightarrow \mathcal{X}_t(T)(\mathbb{Q})_{\text{tors}}$. □

DEFINITION 4.2. For an ABC triple $P = (a,b,c)$, let $F_P = F_P(a,b)$ be the Frey curve given by the Weierstrass model

$$F_P : y^2 = x\,(x-a)\,(x+b)\,.$$

LEMMA 4.3. Let (a,b,c) be an ABC triple which satisfies the assumptions of Lemma 3.2. Then for each T, the Frey curve F_P with $P = (\mathfrak{A}_T, \mathfrak{B}_T, \mathfrak{C}_T)$ has torsion subgroup $F_P(\mathbb{Q})_{\text{tors}} \cong T$.

PROOF. Let $\mathcal{X}_t(T)$ be as defined in Table 1 for $T = C_2 \times C_{2N}$ for $N = 2, 3, 4$ and as defined in (4.1) for $N = 1$. In addition, let u_T, r_T, s_T, w_T, and t_T be as defined in Table 5. We now proceed by cases.

Case I. Suppose $T = C_2 \times C_{2N}$ for $N = 2, 3, 4$. Then the admissible change of variables $x \longmapsto u_T^2 x + r_T$ and $y \longmapsto u_T^3 y + u_T^2 s_T x + w_T$ gives a $\mathbb{Q}$-isomorphism from F_P onto $\mathcal{X}_{t_T}(T)$. In particular, $T \hookrightarrow F_P(\mathbb{Q})_{\text{tors}}$ by Lemma 4.1. By Mazur's Torsion Theorem [**Maz77**] we conclude that $F_P(\mathbb{Q})_{\text{tors}} \cong C_2 \times C_{2N}$ for $N = 3, 4$ and that $F_P(\mathbb{Q})_{\text{tors}}$ is isomorphic to either $C_2 \times C_4$ or $C_2 \times C_8$ if $T = C_2 \times C_4$. For the latter, we observe that our model for $\mathcal{X}_t(T)$ parametrizes elliptic curves E over $\mathbb{Q}(i)$ with $C_4 \times C_4 \hookrightarrow E(\mathbb{Q}(i))_{\text{tors}}$ [**Sil97**, §4]. By Kamienny's Torsion Theorem [**Kam92**] we conclude that $E(\mathbb{Q}(i))_{\text{tors}} \cong C_4 \times C_4$. Thus $\mathcal{X}_t(T)(\mathbb{Q}(i))_{\text{tors}} \cong C_4 \times C_4$ and therefore $C_2 \times C_8 \not\hookrightarrow \mathcal{X}_t(T)(\mathbb{Q}(i))_{\text{tors}}$. Hence $\mathcal{X}_t(T)(\mathbb{Q})_{\text{tors}} \cong C_2 \times C_4$.

Case II. Suppose $T = C_2 \times C_2$ and $T_4 = C_2 \times C_4$. Then there is a 2-isogeny $\phi : \mathcal{X}_t(T_4) \to \mathcal{X}_t(T)$ obtained by applying Vélu's formulas [**V71**] to the elliptic curve $\mathcal{X}_t(T_4)$ and its torsion point $2P$ where $P = (0, 0)$ is the torsion point of order 4 of $\mathcal{X}_t(T_4)$.

Next, observe that via the First Isomorphism Theorem:

(4.2)
$$|\mathcal{X}_t(T)(\mathbb{Q})_{\text{tors}}|\,|\mathcal{X}_t(T_4)(\mathbb{Q})[\phi]| = |\mathcal{X}_t(T_4)(\mathbb{Q})_{\text{tors}}|\,[\mathcal{X}_t(T_4)(\mathbb{Q})_{\text{tors}} : \phi(\mathcal{X}_t(T)(\mathbb{Q})_{\text{tors}})]\,.$$

By Case I above we have that $|\mathcal{X}_t(T_4)(\mathbb{Q})_{\text{tors}}| = 8$ which implies that the only prime dividing $|\mathcal{X}_t(T)(\mathbb{Q})_{\text{tors}}|$ is 2 since ϕ is a 2-isogeny.

Next, we consider the admissible change of variables $x \longmapsto u_T^2 x + r_T$ and $y \longmapsto u_T^3 y + u_T^2 s_T x + w_T$ which gives a $\mathbb{Q}$-isomorphism from F_P onto $\mathcal{X}_{t_T}(T)$. In particular, $C_2 \times C_2 \hookrightarrow F_P(\mathbb{Q})_{\text{tors}}$ by Lemma 4.1. By the proof of Lemma 4.1, $\mathcal{X}_t(T)$ is $\mathbb{Q}$-isomorphic to the elliptic curve given by the Weierstrass model

$$y^2 = x\left(x - 8ab\left(a^2 + b^2\right)\right)\left(x + (a - b)^4\right).$$

This model satisfies the assumptions of [**Ono96**, Main Theorem 1] and therefore we have that $\mathcal{X}_t(T)(\mathbb{Q})_{\text{tors}} \cong C_2 \times C_2$ if $8ab\left(a^2 + b^2\right)$ is not a square. If it were a square we would have a nontrivial integer solution to the Diophantine equation $x^4 - y^4 = z^2$ since

$$8ab\left(a^2 + b^2\right) + (a - b)^4 = (a + b)^4\,.$$

This contradicts Fermat's Theorem and therefore $\mathcal{Y}_t(T)(\mathbb{Q})_{\text{tors}} \cong C_2 \times C_2$. □

THEOREM 4.4. Let $T = C_2 \times C_{2N}$ for $N = 1, 2, 3, 4$ and consider the sequence of good ABC triples P_j^T defined in Proposition 3.4. Then for each $j \geq 1$, the Frey curve $F_{P_j^T}$ determined by P_j^T has torsion subgroup $F_{P_j^T}(\mathbb{Q})_{\text{tors}} \cong C_2 \times C_{2N}$.

PROOF. In Proposition 3.4, we saw that each P_j^T satisfies the assumptions of Lemma 3.2. Consequently, the Theorem follows from Lemma 4.3. □

The case of $N = 2, 4$ in Theorem 4.4 was proven by the author alongside Watts and Tillman [**BTW10**] as part of the Mathematical Sciences Research Institute Undergraduate Program.

5. Examples of good ABC triples

DEFINITION 5.1. For an ABC triple $P = (a, b, c)$, define the quality $q(P)$ of P to be

$$q(P) = \frac{\log(c)}{\log(\mathrm{rad}(abc))}.$$

In particular, P is a good ABC triple is equivalent to $q(P) > 1$.

EXAMPLE 5.2. For $T = C_2 \times C_{2N}$ where $N = 1, 2$ let $P_0 = \left(2^5, 7^2, 3^4\right)$. Then P_0 is a good ABC triple since $q(P) \approx 1.1757$. By Proposition 3.4, this good ABC triple results in two distinct infinite sequences of good ABC triples P_j^T.

For $T = C_2 \times C_6$, let $P_0 = \left(2^4 3^3, 17^3 61, 5^3 7^4\right)$. Then P_0 is a good ABC triple since $q(P) \approx 1.0261$. Moreover, $\frac{17^3 61}{2^4 3^3} > \theta_T$. By Proposition 3.4, this good ABC triple results in an infinite sequence of good ABC triples P_j^T.

For $T = C_2 \times C_8$, let $P_0 = \left(2^2, 11^2, 5^3\right)$. Then P_0 is a good ABC triple since $q(P) \approx 1.0272$. Moreover, $\frac{121}{4} > \theta_T$. By Proposition 3.4, this good ABC triple results in an infinite sequence of good ABC triples P_j^T.

Table 2 gives a_1 and b_1 of $P_j^T = (a_j, b_j, c_j)$ as well as the quality $q\left(P_j^T\right)$ for $j = 1, 2, 3$. We note that the values of a_j and b_j are not given for $j \geq 2$ due to the size of these quantities. For $T = C_2 \times C_{2N}$ for $N = 3, 4$, we only compute $q\left(P_j^T\right)$ for $j = 1, 2$ due to computational limitations.

TABLE 2. Table for Example 5.2

T	a_1	b_1	$q\left(P_1^T\right)$	$q\left(P_2^T\right)$	$q\left(P_3^T\right)$
$C_2 \times C_2$	$2^5 11^2 14657$	$3^8 13^4$	1.0755	1.0324	1.015
$C_2 \times C_4$	$2^{12} 7^4$	$3^8 17^2$	1.2425	1.0531	1.0130
$C_2 \times C_6$	$2^{16} 3^9 17^3 61$	$5^9 7^{12} 11 \cdot 27127$	1.1211	1.0278	–
$C_2 \times C_8$	$2^{12} 11^8$	$7 \cdot 31 \cdot 503 \cdot 1951 \cdot 14657^2$	1.0331	1.0040	–

6. Infinitely many good Frey curves

Recall that the ABC Conjecture is equivalent to the modified Szpiro conjecture which states that for every $\epsilon > 0$ there are finitely many rational elliptic curves E satisfying

$$N_E^{6+\epsilon} < \max\left\{\left|c_4^3\right|, c_6^2\right\}$$

where N_E is the conductor of E and c_4 and c_6 are the invariants associated to a minimal model of E. The following definition gives the analog of good ABC triples and the quality of an ABC triple in the context of elliptic curves.

DEFINITION 6.1. Let E be a rational elliptic curve with minimal discriminant $\Delta_E^{\min}$ and associated invariants c_4 and c_6. Define the **modified Szpiro ratio** $\sigma_m(E)$ and **Szpiro ratio** $\sigma(E)$ of E to be the quantities

$$\sigma_m(E) = \frac{\log \max\left\{\left|c_4^3\right|, c_6^2\right\}}{\log N_E} \qquad \text{and} \qquad \sigma(E) = \frac{\log\left|\Delta_E^{\min}\right|}{\log N_E}$$

where N_E is the conductor of E. We say that E is **good** if $\sigma_m(E) > 6$.

Let $P = (a, b, c)$ be an ABC triple with a even and $b \equiv 1 \bmod 4$. For $T = C_2 \times C_{2N}$ where $N = 1, 2, 3, 4$, let $\mathfrak{A}_T = \mathfrak{A}_T(a, b)$, $\mathfrak{B}_T = \mathfrak{B}_T(a, b)$, $\mathfrak{C}_T = C_T(a, b)$, and $\mathfrak{D}_T = \mathfrak{D}_T(a, b)$ be as defined in Table 6. Assume further that $a \equiv 0 \bmod 3$ if $T = C_2 \times C_6$. Then the elliptic curve $F_T = F_T(a, b)$ given by the Weierstrass model

$$F_T : y^2 = x\,(x - \mathfrak{A}_T)(x + \mathfrak{B}_T)$$

satisfies $F_T(\mathbb{Q})_{\text{tors}} \cong T$ by Lemma 4.3. Moreover, the congruences on $\mathfrak{A}_T$ and $\mathfrak{B}_T$ imply that the Frey curve F_T is semistable with minimal discriminant $\Delta_T = \left(16^{-1}\mathfrak{A}_T\mathfrak{B}_T\mathfrak{C}_T\right)^2$ [**Sil09**, Exercise 8.23]. Consequently, the conductor N_T of F_T satisfies $N_T = \operatorname{rad}(\Delta_T) < |\mathfrak{D}_T|$ and the invariant $c_{4,T} = c_{4,T}(a, b)$ associated with a global minimal model of F_T is as given in Table 3.

Table 3: The Invariant c_4 of F_T

$c_{4,T}$	T
$a^8 + 60a^6b^2 + 134a^4b^4 + 60a^2b^6 + b^8$	$C_2 \times C_2$
$a^8 + 14a^4b^4 + b^8$	$C_2 \times C_4$
$9a^8 + 228a^6b^2 + 30a^4b^4 - 12a^2b^6 + b^8$	$C_2 \times C_6$
$a^{16} - 8a^{14}b^2 + 12a^{12}b^4 + 8a^{10}b^6 + 230a^8b^8 + 8a^6b^{10} + 12a^4b^{12} - 8a^2b^{14} + b^{16}$	$C_2 \times C_8$

LEMMA 6.2. Let $P = (a, b, c)$ be a good ABC triple satisfying $a \equiv 0 \bmod 2$, $b \equiv 1 \bmod 4$, and $\frac{b}{a} > \theta_T$ where θ_T is as given in Lemma 2.1. Assume further that $a \equiv 0 \bmod 3$ if $T = C_2 \times C_6$. Then the Frey curve $F_T = F_T(\mathfrak{A}_T, \mathfrak{B}_T)$ is good and $F_T(\mathbb{Q})_{\text{tors}} \cong T$.

PROOF. By Lemma 4.3, $F_T(\mathbb{Q})_{\text{tors}} \cong T$. Since F_T is a Frey curve we have that the invariants c_4 and c_6 associated to a global minimal model of F_T satisfy $\max\{|c_4^3|, c_6^2\} = c_4^3$ since c_4 and $\Delta_{F_T}^{\min}$ are always positive [**Sil09**, Lemma VIII.11.3]. The congruences on a and b imply that $c_4 = c_{4,T}$. It, therefore, suffices to show that $c_{4,T}^3 - N_T^6 > 0$ where N_T is the conductor of F_T. Since F_T is semistable,

$$N_T = \operatorname{rad}(\mathfrak{A}_T\mathfrak{B}_T\mathfrak{C}_T) < \mathfrak{D}_T$$

by Lemma 3.3. Note that $\mathfrak{D}_T$ is positive since $\frac{b}{a} > \theta_T$. Thus

$$\frac{c_{4,T}^3 - N_T^6}{\mathfrak{D}_T(1,t)^6} > \frac{c_{4,T}(1,t)^3 - \mathfrak{D}_T(1,t)^6}{\mathfrak{D}_T(1,t)^6} \quad \text{for } t = \frac{b}{a} \tag{6.1}$$

Lastly, for each T, the polynomial $c_{4,T}(1,t)^3 - \mathfrak{D}_T(1,t)^6$ is positive on the open interval (θ_T, ∞) from which we conclude that F_T is a good elliptic curve. □

THEOREM 6.3. For each T, let $P_0^T = (a_0, b_0, c_0)$ be a good ABC triple satisfying $a_0 \equiv 0 \bmod 2$, $b_0 \equiv 1 \bmod 4$, and $\frac{b_0}{a_0} > \theta_T$ where θ_T is as given in Lemma 2.1. Assume further that $a_0 \equiv 0 \bmod 3$ if $T = C_2 \times C_6$. For $j \geq 1$, define P_j^T recursively by

$$P_j^T = (a_j, b_j, c_j) = (\mathfrak{A}_T(a_{j-1}, b_{j-1}), \mathfrak{B}_T(a_{j-1}, b_{j-1}), \mathfrak{C}_T(a_{j-1}, b_{j-1})).$$

Then for each j, the Frey curve $F_T(a_j, b_j)$ is good and $F_T(a_j, b_j)(\mathbb{Q})_{\text{tors}} \cong T$.

PROOF. By Proposition 3.4, $P_j^T = (a_j, b_j, c_j)$ satisfies $a_j \equiv 0 \bmod 2$, $b_j \equiv 1 \bmod 4$, and $\frac{b_j}{a_j} > \theta_T$ for each j. For $T = C_2 \times C_6$, if $a_0 \equiv 0 \bmod 3$, then

$a_j \equiv 0 \mod 3$ for each j. Hence P_j^T is a good ABC triple for each j by Proposition 3.4. Therefore the result follows by Lemma 6.2. □

In Example 5.2 we began with a good ABC triple $P_0 = (a_0, b_0, c_0)$. For each T, we constructed an infinite sequence of good ABC triples $P_j^T = (a_j, b_j, c_j)$. By Theorem 6.3, each Frey curve $F_T(a_j, b_j)(\mathbb{Q})_{\text{tors}}$ is a good elliptic curve with torsion subgroup isomorphic to T. Table 4 lists the modified Szpiro ratios of the Frey curves corresponding to P_j^T. Due to computational limitations, we could only compute these ratios up to $j = 3$.

TABLE 4. Example of Good Frey Curves

T	$C_2 \times C_2$	$C_2 \times C_4$	$C_2 \times C_6$	$C_2 \times C_8$
$\sigma_m(F_T(a_1, b_1))$	6.4204	7.4219	6.7269	6.1985
$\sigma_m(F_T(a_2, b_2))$	6.1912	6.3124	6.1666	6.0241
$\sigma_m(F_T(a_3, b_3))$	6.0901	6.0769		

7. Table of polynomials

TABLE 5. Admissible Change of Variables for Lemma 4.3

T	u_T	r_T	s_T	w_T	t_T
$C_2 \times C_2$	a^2	0	0	0	$\frac{b}{a}$
$C_2 \times C_4$	$2(a-b)^2$	$-2ab(a-b)^2$	$(a-b)^2$	$-2ab(a-b)^2(a^2+b^2)$	$\frac{b}{a}$
$C_2 \times C_6$	$9a^2-b^2$	$-4a^2(a+b)(-3a+b)$	$5a^2-b^2$	$36a^6-40a^4b^2+4a^2b^4$	$\frac{9a+b}{a+b}$
$C_2 \times C_8$	$\frac{1}{2a(a+b)(b^2-2ab-a^2)}$	$\frac{ab(a^2+b^2)}{(a+b)^2(b^2-2ab-a^2)}$	$\frac{a^4+4a^3b-b^4}{2a(a+b)(b^2-2ab-a^2)}$	$\frac{ab^2(a^2+b^2)^2}{(a+b)^3(b^2-2ab-a^2)^2}$	$\frac{a}{2(b-a)}$

TABLE 6. Polynomials and Rational Functions

T	$C_2 \times C_2$	$C_2 \times C_4$	$C_2 \times C_6$	$C_2 \times C_8$
$\mathfrak{A}_T$	$8ab\left(a^2+b^2\right)$	$(2ab)^2$	$16a^3b$	$(2ab)^4$
$\mathfrak{B}_T$	$(a-b)^4$	$\left(a^2-b^2\right)^2$	$(a+b)^3(b-3a)$	$\left(a^4-6a^2b^2+b^4\right)\left(a^2+b^2\right)^2$
$\mathfrak{C}_T$	$(a+b)^4$	$\left(a^2+b^2\right)^2$	$(3a+b)(b-a)^3$	$\left(a^2-b^2\right)^4$
$\mathfrak{D}_T$	b^4-a^4	b^4-a^4	$\left(b^2-a^2\right)\left(b^2-9a^2\right)$	$\left(a^4-6a^2b^2+b^4\right)\left(b^4-a^4\right)$
$\mathfrak{A}_T^r$	$ab\left(a^2+b^2\right)$	ab	ab	ab
$\mathfrak{B}_T^r$	$(a-b)$	a^2-b^2	$(a+b)(b-3a)$	$\left(a^4-6a^2b^2+b^4\right)\left(a^2+b^2\right)$
$\mathfrak{C}_T^r$	$a+b$	a^2+b^2	$(3a+b)(b-a)$	a^2-b^2
f_T	$\frac{(1-t)^4}{8t(1+t^2)}-t$	$\frac{\left(1-t^2\right)^2}{(2t)^2}-t$	$\frac{(1+t)^3(t-3)}{16t}-t$	$\frac{\left(1-6t^2+t^4\right)\left(1+t^2\right)^2}{(2t)^4}-t$
g_T	$4t^3+6t^2+4t+2$	$2t^2+2$	$4t^2+8t-12$	$2t^6+6t^4-10t^2+2$
θ_T	1	1	4.87517	3.17374
U_T	$5a^3r+20a^2br+29ab^2r+$ $16b^3r+16a^3s+29a^2bs+$ $20ab^2s+5b^3s$	a^2r+2b^2r+ $2a^2s+b^2s$	$-54a^3r+144a^2br-117ab^2r+$ $24b^3r-8a^3s+$ $6a^2bs-b^3s$	$4a^6r-15a^4b^2r+20a^2b^4r-$ $10b^6r-10a^6s+$ $20a^4b^2s-15a^2b^4s+4b^6s$
V_T	$-5a^3r+20a^2br-29ab^2r+$ $16b^3r+16a^3s-$ $29a^2bs+20ab^2s-5b^3s$	$-a^2r+2b^2r+$ $2a^2s-b^2s$	$54a^3r+144a^2br+$ $117ab^2r+24b^3r-$ $8a^3s-6a^2bs+b^3s$	$-4a^6r+15a^4b^2r+44a^2b^4r+$ $26b^6r+26a^6s+$ $44a^4b^2s+15a^2b^4s-4b^6s$
W_T	$32\left(ra^7+sb^7\right)$	$4\left(ra^6+sb^6\right)$	$48\left(ra^7+sb^7\right)$	$16\left(ra^{14}+sb^{14}\right)$

References

[BTW10] Alexander J. Barrios, Caleb Tillman, and Charles Watts, *Exceptional abc triples for frey curves with torsion subgroups* $z_2 \times z_4$ *and* $z_2 \times z_8$, MSRI Journal (2010).

[CR01] Brian Conrad and Karl Rubin (eds.), *Arithmetic algebraic geometry*, IAS/Park City Mathematics Series, vol. 9, American Mathematical Society, Providence, RI; Institute for Advanced Study (IAS), Princeton, NJ, 2001. Including papers from the Graduate Summer School of the Institute for Advanced Study/Park City Mathematics Institute held in Park City, UT, June 20–July 10, 1999. MR1860012

[HLP00] Everett W. Howe, Franck Leprévost, and Bjorn Poonen, *Large torsion subgroups of split Jacobians of curves of genus two or three*, Forum Math. **12** (2000), no. 3, 315–364, DOI 10.1515/form.2000.008. MR1748483

[Kam92] S. Kamienny, *Torsion points on elliptic curves and q-coefficients of modular forms*, Invent. Math. **109** (1992), no. 2, 221–229, DOI 10.1007/BF01232025. MR1172689

[Kub76] Daniel Sion Kubert, *Universal bounds on the torsion of elliptic curves*, Proc. London Math. Soc. (3) **33** (1976), no. 2, 193–237, DOI 10.1112/plms/s3-33.2.193. MR0434947

[Mas90] D. W. Masser, *Note on a conjecture of Szpiro*, Astérisque **183** (1990), 19–23. Séminaire sur les Pinceaux de Courbes Elliptiques (Paris, 1988). MR1065152

[Maz77] B. Mazur, *Modular curves and the Eisenstein ideal*, Inst. Hautes Études Sci. Publ. Math. **47** (1977), 33–186 (1978). With an appendix by Mazur and M. Rapoport. MR488287

[Oes88] Joseph Oesterlé, *Nouvelles approches du "théorème" de Fermat* (French), Astérisque **161-162** (1988), Exp. No. 694, 4, 165–186 (1989). Séminaire Bourbaki, Vol. 1987/88. MR992208

[Ono96] Ken Ono, *Euler's concordant forms*, Acta Arith. **78** (1996), no. 2, 101–123, DOI 10.4064/aa-78-2-101-123. MR1424534

[Sil97] Alice Silverberg, *Explicit families of elliptic curves with prescribed mod N representations*, Modular forms and Fermat's last theorem (Boston, MA, 1995), Springer, New York, 1997, pp. 447–461. MR1638488

[Sil09] Joseph H. Silverman, *The arithmetic of elliptic curves*, 2nd ed., Graduate Texts in Mathematics, vol. 106, Springer, Dordrecht, 2009. MR2514094

[V71] Jacques Vélu, *Isogénies entre courbes elliptiques* (French), C. R. Acad. Sci. Paris Sér. A-B **273** (1971), A238–A241. MR294345

Department of Mathematics and Statistics, Carleton College, Northfield, Minnesota 55057

Email address: abarrios@carleton.edu

Contemporary Mathematics
Volume **759**, 2020
https://doi.org/10.1090/conm/759/15266

On the existence of $C^{(n)}$-almost automorphic mild solutions of certain differential equations in Banach spaces

Gaston M. N'Guérékata and Gisèle Mophou

Abstract. In this paper we study the existence and uniqueness of almost automorphic mild solutions to the semilinear equation $u'(t) = A(t)u(t) + f(t, u(t))$, $t \in \mathbb{R}$ where $A(t)$ satisfies the Acquistapace-Terreni conditions and generates an exponentially stable family of operators $(U(t,s)_{t\geq s}$ and $f(t,u)$ is almost automorphic in $t \in \mathbb{R}$ for any $u \in X$. This extends a well-known result by G.M. N'Guérékata in the case where $A(t) = A$ does not depend on t. We also investigate the existence of mild C^n-almost automorphic solutions of the linear version of the above equation when A is an operator of simplest type.

1. Introduction

The concept of almost automorphic functions was introduced in the literature by S. Bochner [**4, 5**] while studying some problems in differential geometry. It was then developed by W.A. Veech [**17**] for functions on groups and M. Zaki [**18**] for functions on the real line. Further developments were boosted by G.M. N'Guérékata's books [**13, 15**]. The concept turns out to generalize the concept of almost periodic functions in the sense of Bohr and has applications in several problems in mathematics and the sciences (cf. for instance [**6, 8–12, 19, 21**] and references therein).

In [**14**], the author studied the semilinear equation in a Banach space X

$$u'(t) = Au(t) + f(t, u(t)), \quad t \in \mathbb{R} \tag{1.1}$$

where $A : D(A) \subset X \to X$ is the infinitesimal generator of an exponentially stable C_0-semigroup of bounded operators, and $f(t, u)$ is almost automorphic in t for any $u \in X$ and is lipschitzian in t uniformly in u. He proved that the equation possesses a unique almost automorphic mild solution. This result has been extended to general cases where f is not necessarily lipschitzian and the operator A not necessarily a generator of a C_0-semigroup .

2020 *Mathematics Subject Classification.* 34G10, 47D06, 45M05.

This work has been conducted during the first author's visit at AIMS-Cameroon Research Center in Limbe, Cameroon in March 2019. He is grateful for the invitation by the second author, supported by the Alexander von Humboldt foundation, under the program financed by the BMBF entitled "German Research Chairs".

In [**3**], the authors studied the existence of S-asymptotically ω-periodic mild solution of the nonautonomous semilinear equation

$$u'(t) = A(t)u(t) + f(t, u(t)), \quad t \in \mathbb{R}^+ \tag{1.2}$$

where $A(t)$ generates a ω-periodic exponentially stable evolutionary process $(U(t,s)_{t\geq s}$.

Motivated by the above results, we prove in the present paper that Equation(1.2) considering $A(t)$ generates an evolutionary process $(U(t,s)_{t\geq s}$ which is not necessarily ω-periodic, admits a unique almost automorphic mild solution assuming appropriate conditions on the function $f(t,u)$. This is our main result Theorem 4.3. We also discuss C^n-almost automorphic mild solutions to the linear version of the above Equation (1.1). These results extend the ones obtained in [**2**] to the case of this type of functions and the operator A being of simplest type.

2. $C^{(n)}$-almost automorphic functions

DEFINITION 2.1. [**4, 5, 16, 18**] A continuous function $f : \mathbb{R} \to X$ is said to be almost automorphic if for every sequence or real numbers (s'_n) there exists a subsequence (s_n) such that

$$\lim_{n\to\infty} f(t+s_n) = g(t)$$

exists for each $t \in \mathbb{R}$ and

$$\lim_{n\to\infty} g(t-s_n) = f(t)$$

for each $t \in \mathbb{R}$.

REMARK 2.2. (i) The function g in the definition above is measurable, but not necessarily continuous.
(ii) If the convergence above is uniform in $t \in \mathbb{R}$, then f is almost periodic in the sense of Bohr.

PROPOSITION 2.3. [**13**] If f, f_1, f_2 are almost automorphic functions $\mathbb{R} \to X$, λ a scalar, then the following are true:

i) $f_1 + f_2,\ \lambda f$ are almost automorhic.
ii) the translation $f_\tau(t) := f(t+\tau),\ \ t \in \mathbb{R}$ is almost automorphic.
iii) the function $\tilde{f}(t) := f(-t),\ \ t \in \mathbb{R}$ is almost automorphic.
iv) the range of f $\mathcal{R}_f := \{f(t) :\ t \in \mathbb{R}\}$ is relatively compact in X.

THEOREM 2.4. [**13**] *If (f_n) is a sequence of almost automorphic functions which convergent uniformly in $t \in \mathbb{R}$ to f, then f is almost automorphic.*

REMARK 2.5. Denote $AA(X)$ the space of all almost automorphic functions $\mathbb{R} \to X$. Equipped with the norm $\|f\| = \sup_{\mathbb{R}} \|f(t)\|$, $AA(X)$ turns out to be a Banach space.
If we denote by $AP(X)$ the Banach space of all almost periodic functions in the sens of Bohr, then we have

$$AP(X) \subset AA(X).$$

The inclusion is strict. Indeed the function

$$f(t) = sin(\frac{1}{2 + cost + cos\sqrt{2}t}) \tag{2.1}$$

is almost automorphic but not almost periodic.

We recall the following Zheng-Ding-N'Guérékata theorem in [**21**] which provides an information on the "amount" of almost automorphic functions which are not almost periodic. We also give a different and more elegant proof than in [**21**] suggested by Marek Balcerzak in his review MR3036705.

THEOREM 2.6. *$AP(X)$ is a set of first category in $AA(X)$.*

PROOF. It suffices to note from the above that $AP(X)$ is a proper closed subspace of $AA(X)$ equipped with the supnorm. Therefore it is of first category in $AA(X)$. □

DEFINITION 2.7. [**9**] A continuous function $f : \mathbb{R} \to X$ is said to be C^n-almost automorphic for $n \geq 1$ if for $i = 1, 2, ..., n$, the i-th derivative $f^{(i)}$ of f is almost automorphic.

PROPOSITION 2.8. [**10, 14**] The set $AA^{(n)}(X)$ of all C^n-almost automorphic functions $f : \mathbb{R} \to X$ is a Banach space under the norm

$$\|f\|_n = \sup_{t\in\mathbb{R}}(\sum_{i=1}^{n} \|f^{(i)}(t)\|).$$

THEOREM 2.9. *Suppose that X does not contain a subspace isomorphic to c_o and $f \in AA^{(n)}(X)$. Then*

$$F(t) := \int_0^t f(\xi)d\xi \in AA^{(n+1)}(X) \text{ iff the } \mathcal{R}_f := \{f(t) : t \in \mathbb{R}\} \text{ is bounded in} X.$$

PROOF. The "if" part is immediate. Let's prove the "only" part. We use the induction principle. The case $n = 0$ is well-known (cf. for instance [**14**]). Assume that $f \in AA^{(k-1)}(X)$ for some k. That is $F' = f \in AA^{(k-1)}(X)$. But by assumption f is bounded. Therefore $F \in AA^{(k)}(X)$. The proof is complete. □

3. Linear Equations

We will consider in this section, the following equation in a complex Banach space X which does not contain a subspace isomorphic to c_o.

$$u'(t) = Au(t) + f(t), \quad t \in \mathbb{R}, \tag{3.1}$$

where $A : D(A) \subset X \to X$ is a linear operator and $f \in C(\mathbb{R}, X)$.
In what follows we will use the notation $\Pi := \{z \in \mathbb{C} : \mathcal{R}ez \neq 0\}$

DEFINITION 3.1. [**7, 20**] A linear operator $A : D(A) \subset X \to X$ is said to be of simplest type if $A \in B(X)$ and

$$A = \sum_{i=1}^{n} \lambda_i P_i,$$

where $\lambda_i \in \mathbb{C}, i = 1, 2, ...n$ and $(P_i)_{1\leq i\leq n}$ forms a complete system $\sum_{i=1}^{n} P_i = I$ of mutually disjoint operators on X, i.e. $P_iP_j = \delta_{ij}P_j$

LEMMA 3.2. *Let's consider Equation(3.1) with $A = \lambda \in \mathbb{C}$ and $f \in AA^{(n)}(X)$. Then every bounded solution u satisfies*

$$u \in AA^{(n)}(X), \quad \lambda \in \Pi$$

and

$$u \in AA^{(n+1)}(X), \quad \lambda \notin \Pi.$$

PROOF. The proof is an immediate consequence of Lemma 4.1 in [**9**] □

THEOREM 3.3. *Assume that $f \in AA^{(n)}(X)$ and A is of simplest type. Then every bounded solution u of Eq.(3.1) satisfies*

$$u \in AA^{(n)}(X), \quad \lambda \in \Pi$$

and

$$u \in AA^{(n+1)}(X), \quad \lambda \notin \Pi.$$

PROOF. Applying the projection P_k to Eq.(3.1) gives

$$P_k u'(t) = \frac{d}{dt}(P_k u)(t) = P_k(\sum_{i=1}^{n} \lambda_i P_i) u(t) + P_k f(t) = \lambda_k (P_k u)(t) + (P_k f)(t)$$

Observe that $P_k f \in AA^{(n)}$ since P_k is a bounded linear operator. Therefore by the Lemma 3.2 above,

$$P_k u \in AA^{(n)}(X), \quad \lambda \in \Pi$$

and

$$P_k u \in AA^{(n+1)}(X), \quad \lambda \notin \Pi.$$

Now since $u(t) = \sum_{i=1}^{n} (P_i u)(t)$, the conclusion follows immediately. □

4. Nonautonomous case

Consider now in a complex Banach space X the equation

$$u'(t) = A(t)u(u) + f(t), \quad t \in \mathbb{R}. \tag{4.1}$$

And assume that $A(t),\ t \in \mathbb{R}$, satisfies the Acquistapace-Terreni conditions [**1**]. That means there exist constants $\lambda_0 \geq 0,\ \theta \in (\frac{\pi}{2}, \pi),\ L, K \geq 0$ and $\alpha, \beta \in (0, 1]$ with $\alpha + \beta > 1$ such that

$$\sum_{\theta} \cup \{0\} \subset \rho(A(t) - \lambda_0), \quad \|R(\lambda, A(t) - \lambda_0)\| \leq \frac{K}{1 + |\lambda|} \tag{4.2}$$

and

$$\|(A(t) - \lambda_0) R(\lambda, A(t) - \lambda_0)[R(\lambda_0, A(t) - R(\lambda_0, A(s))]\| \leq L|t - s|^{\alpha} |\lambda|^{\beta} \tag{4.3}$$

for $t, s \in \mathbb{R},\ \ \lambda \in \sum_{\theta} := \{\lambda \in \mathbb{C}^{\star} : |arg\lambda| \leq \theta\}$. Then from [**1**] Theorem 2.3, there exists a unique evolution family $\{U(t,s)\}_{-\infty < s \leq t < \infty}$ on X which governs the linear version of (4.1).

The family $\{U(t,s)\}_{-\infty < s \leq t < \infty}$ satisfies the following properties

(i) $U(t,t) = I, \ \ \forall t \in \mathbb{R}$,
(ii) $U(t,r)T(r,s) = U(t,s), \ \ \forall t \geq r \geq s$,
(iii) The map $(t,s) \to U(t,s)\xi$ is continuous for every $\xi \in X$.

DEFINITION 4.1. A function $u \in C(\mathbb{R}, X)$ is said to be a mild solution to Eq.(4.1) if it has the integral representation

$$u(t) = U(t,s)u(s) + \int_s^t U(t,s)f(s)ds, \quad t, s \in \mathbb{R},\ t \geq s. \tag{4.4}$$

Under the above assumptions, it can be proved as in G. N'Guérékata et *al.* [**2**], Theorem 3.6 that any mild solution can be written as

$$u(t) = \int_{-\infty}^{t} U(t,s)f(s)ds, \quad t \in \mathbb{R} \tag{4.5}$$

Let make the following assumptions:

H1. $f \in AA(X)$

H2. There exists a function $\varphi \in L^1(\mathbb{R}, \mathbb{R}^+)$ such that

$$\|U(t,s)\| \leq \varphi(t-s).$$

THEOREM 4.2. *Assume that assumptions (H1)-(H2) hold true. Assume also that that $A(t)$, $t \in \mathbb{R}$, satisfies the Acquistapace-Terreni conditions.*

Then Eq.(4.1) admits a unique mild solution in $AA(X)$.

PROOF. Observing that $\lim_{t\to\infty} \varphi(t) = 0$, it can be shown as in [**2**] that the function $\Lambda : \mathbb{R} \to BC(\mathbb{R}, X)$ defined by

$$\Lambda(t) := \int_{-\infty}^{t} U(t,s)f(s)ds$$

is the unique solution to Eq.(4.1). Now let's prove it is almost automorphic.

Let (s_n) be an arbitrary sequence of real numbers. Since f is almost automorphic, there exists a subsequence (s_n) of (s'_n) such that

$$\lim_{n\to\infty} f(t+s_n) = g(t)$$

exists for each $t \in \mathbb{R}$ and

$$\lim_{n\to\infty} g(t-s_n) = f(t)$$

for each $t \in \mathbb{R}$.

Consider $\tilde{\Lambda} : \mathbb{R} \to BC(\mathbb{R}, X)$ such that

$$\tilde{\Lambda}(t) := \int_{-\infty}^{t} U(t,s)g(s)ds.$$

In view of the properties of U and g, this functions is well defined. Now we have

$$\begin{aligned} \Lambda(t+s_n) - \tilde{\Lambda}(t) &= \int_{-\infty}^{t+s_n} U(t+s_n,s)f(s)ds - \int_{-\infty}^{t} U(t,s)g(s)ds \\ &= \int_{-\infty}^{t} U(t+s_n, s+s_n)f(s+s_n)ds - \int_{-\infty}^{t} U(t,s)g(s)ds. \end{aligned}$$

Now observe that for any $n = 1, 2, \ldots$ we have

$$\|U(t+s_n, s+s_n)f(s+s_n)\| \leq M\varphi(t-s)\|f\|_\infty$$

and the right hand member of the inequality belongs to $L^1(\mathbb{R}, \mathbb{R}^+)$. Therefore, in virtue of the Lebesgue Dominated Convergence theorem,

$$\int_{-\infty}^{t} U(t+s_n, s+s_n)f(s+s_n)ds \to \int_{-\infty}^{t} U(t,s)g(s)ds, \text{ as } n \to \infty.$$

In other words,

$$\Lambda(t+s_n) \to \tilde{\Lambda}(t), \text{ as } n \to \infty.$$

Analogously we can prove that

$$\tilde{\Lambda}(t - s_n) \to \Lambda(t), \text{ as } n \to \infty.$$

That means $\Lambda(t) \in AA(X)$.

□

Consider now in a complex Banach space X the semilinear equation

$$u'(t) = A(t)u(u) + F(t, u(t)), \quad t \in \mathbb{R}. \tag{4.6}$$

THEOREM 4.3. *And assume that $A(t)$, $t \in \mathbb{R}$, satisfies the Acquistapace-Terreni conditions and (H2) holds. Assume also that $F(t,x)$ is almost automorphic in $t \in \mathbb{R}$ uniformly in $x \in K$ for any bounded set $K \subset X$ and*

$$\|F(t,x) - F(t,y)\| \leq L\|x - y\|, \quad t \in \mathbb{R}, \ x, y \in X.$$

Then Eq.(4.6) has a unique mild solution in $AA(X)$ if we have $L\|\varphi\|_{L^1(\mathbb{R},\mathbb{R}^+)} < 1$ in addition.

PROOF. Consider the operator $\Upsilon : AA(X) \to AA(X)$ defined by

$$(\Upsilon u)(t) := \int_{-\infty}^{t} U(t,s)F(s,u(s))ds.$$

In view of Theorem 2.2.5 [**13**] and 4.2 above this operator is well-defined.

Now take $u, v \in AA(X)$. Then we get

$$\begin{aligned} \|(\Upsilon u)(t) - (\Upsilon v)(t)\| &= \left\| \int_{-\infty}^{t} U(t,s)[F(s,u(s)) - F(s,v(s))ds \right\| \\ &\leq \int_{-\infty}^{t} \|U(t,s)\| \|F(s,u(s)) - F(s,v(s))\| ds \\ &\leq L\|\varphi\|_{L^1(\mathbb{R},\mathbb{R}^+)} \|u - v\|_\infty. \end{aligned}$$

Therefore

$$\|(\Upsilon u) - (\Upsilon v)\|_\infty \leq L\|\varphi\|_{L^1(\mathbb{R},\mathbb{R}^+)} \|u - v\|_\infty$$

and the conclusion follows in virtue of the Banach fixed point theorem. □

References

[1] Paolo Acquistapace, *Evolution operators and strong solutions of abstract linear parabolic equations*, Differential Integral Equations **1** (1988), no. 4, 433–457. MR945820

[2] Jean-Bernard Baillon, Joël Blot, Gaston M. N'Guérékata, and Denis Pennequin, *On $C^{(n)}$-almost periodic solutions to some nonautonomous differential equations in Banach spaces*, Comment. Math. (Prace Mat.) **46** (2006), no. 2, 263–273. MR2287690

[3] Joel Blot, Philippe Cieutat, and Gaston M. N'Guérékata, *S-asymptotically ω-periodic functions and applications to evolution equations*, Afr. Diaspora J. Math. **12** (2011), no. 2, 113–121. MR2847308

[4] S. Bochner, *Uniform convergence of monotone sequences of functions*, Proc. Nat. Acad. Sci. U.S.A. **47** (1961), 582–585, DOI 10.1073/pnas.47.4.582. MR126094

[5] S. Bochner, *A new approach to almost periodicity*, Proc. Nat. Acad. Sci. U.S.A. **48** (1962), 2039–2043, DOI 10.1073/pnas.48.12.2039. MR145283

[6] Dariusz Bugajewski and Toka Diagana, *Almost automorphy of the convolution operator and applications to differential and functional differential equations*, Nonlinear Stud. **13** (2006), no. 2, 129–140. MR2228933

[7] Ju. L. Dalec′kiĭ and M. G. Kreĭn, *Stability of solutions of differential equations in Banach space*, American Mathematical Society, Providence, R.I., 1974. Translated from the Russian by S. Smith; Translations of Mathematical Monographs, Vol. 43. MR0352639

[8] Brahim Es-sebbar, Khalil Ezzinbi, and Gaston M. N'Guérékata, *Bohr-Neugebauer property for almost automorphic partial functional differential equations*, Appl. Anal. **98** (2019), no. 1-2, 381–407, DOI 10.1080/00036811.2017.1382686. MR3902137

[9] Khalil Ezzinbi, Samir Fatajou, and Gaston Mandata N'Guérékata, *C^n-almost automorphic solutions for partial neutral functional differential equations*, Appl. Anal. **86** (2007), no. 9, 1127–1146, DOI 10.1080/00036810701584591. MR2351637

[10] Khalil Ezzinbi, Valerie Nelson, and Gaston N'Guérékata, *$C^{(n)}$-almost automorphic solutions of some nonautonomous differential equations* (English, with English and Spanish summaries), Cubo **10** (2008), no. 2, 61–74. MR2431285

[11] Marko Kostić, *Generalized almost automorphic and generalized asymptotically almost automorphic solutions of abstract Volterra integro-differential inclusions*, Fract. Differ. Calc. **8** (2018), no. 2, 255–284, DOI 10.7153/fdc-2018-08-16. MR3891977

[12] Marko Kostić, *On generalized $C^{(n)}$-almost periodic solutions of abstract Volterra integro-differential equations*, Novi Sad J. Math. **48** (2018), no. 1, 73–91, DOI 10.30755/nsjom.06635. MR3825369

[13] Gaston M. N'Guérékata, *Almost automorphic and almost periodic functions in abstract spaces*, Kluwer Academic/Plenum Publishers, New York, 2001. MR1880351

[14] Gaston M. N'Guérékata, *Existence and uniqueness of almost automorphic mild solutions to some semilinear abstract differential equations*, Semigroup Forum **69** (2004), no. 1, 80–86, DOI 10.1007/s00233-003-0021-0. MR2063980

[15] Gaston M. N'Guérékata, *Topics in almost automorphy*, Springer-Verlag, New York, 2005. MR2107829

[16] Gaston Mandata N'Guérékata, *Spectral theory for bounded functions and applications to evolution equations*, Mathematics Research Developments, Nova Science Publishers, Inc., New York, 2017. MR3676744

[17] W. A. Veech, *Almost automorphic functions on groups*, Amer. J. Math. **87** (1965), 719–751, DOI 10.2307/2373071. MR187014

[18] M. Zaki, *Almost automorphic solutions of certain abstract differential equations*, Ann. Mat. Pura Appl. (4) **101** (1974), 91–114, DOI 10.1007/BF02417100. MR364805

[19] S. Zaidman, *Almost-periodic functions in abstract spaces*, Research Notes in Mathematics, vol. 126, Pitman (Advanced Publishing Program), Boston, MA, 1985. MR790316

[20] S. Zaidman, *Topics in abstract differential equations. II*, Pitman Research Notes in Mathematics Series, vol. 321, Longman Scientific & Technical, Harlow; copublished in the United States with John Wiley & Sons, Inc., New York, 1995. MR1327278

[21] Zhe-Ming Zheng, Hui-Sheng Ding, and Gaston M. N'Guérékata, *The space of continuous periodic functions is a set of first category in $AP(X)$*, J. Funct. Spaces Appl., posted on 2013, Art. ID 275702, 3, DOI 10.1155/2013/275702. MR3036705

Morgan State University, Baltimore, MD 21251, U.S.A.
Email address: gaston.N'Guerekata@morgan.edu

Laboratoire LAMIA, Université des Antilles, Campus Fouillole, 97159 Pointe-À-Pitre Guadeloupe (FWI)- Laboratoire MAINEGE, Université Ouaga 3S, 06 BP 10347 Ouagadougou 06, Burkina Faso

Current address: African Institute for Mathematical Sciences (AIMS), P.O. Box 608, Limbe Crystal Gardens, South West Region, Cameroon

Email address: gisele.mophou@univ-antilles.fr

Contemporary Mathematics
Volume **759**, 2020
https://doi.org/10.1090/conm/759/15267

A first hitting time approach to finding effective spreaders in a network

Fern Y. Hunt

ABSTRACT. In a model of network communication based on a random walk in an undirected graph, effective spreaders are a subset of nodes that enable the fastest spread of information. In this setting, an optimal set A minimizes the sum of the expected first hitting times $F(A)$, of random walks that start at nodes outside the set. Here we discuss a polynomial time algorithm for finding or approximating such sets and using recent results from the mixing theory of Markov chains, show some of their structural characteristics.

1. Introduction

This paper is a graph theoretical treatment of communication in a social network. Here a graph is a set of of vertices or nodes denoted by V that are connected to each other by edges E. It is assumed that the graph is connected. In this setting an individual or agent in a social network is identified with a node and the connections with other individuals in the network are represented by edges between the nodes. The formation of a consensus of opinion in a network has been the subject of intense research since the introduction of the DeGroot model in 1974 [**16**]. In the simplest formulation, individual agents synchronously update their function value by averaging it with that of their neighbors. This process is repeated and the resulting function values converge to a single equilibrium thus modelling the evolution of agent opinions to consensus. More sophisticated variations have been studied e.g. allowing for asynchronous communication [**8**], random and switching network topologies, and time varying averaging schemes [**38**]. There has also been work on other models of communication e.g. the voter model [**1**] [**41**], and indirect communication where agents imitate the strategy of their neighbors in a game theoretic context [**19**]

In this paper, we will focus on the case where a subset of agents retain their initial values throughout the process. In the area of opinion dynamics, they are referred to as "stubborn agents". As shown e.g. in [**18**],[**14**], the equilibrium or consensus value of each node is a convex combination of the initial values of the stubborn nodes.

In our discussion, the dynamics of the consensus process will not play a role here since it will always be assumed (for simplicity) that the equilibrium value is

2020 *Mathematics Subject Classification.* Primary 05C69, 60J10; Secondary 90C27, 68R10.
Key words and phrases. Consensus process, first hitting time.

the same for each agent and hence all agents will eventually have the same opinion. Rather, we will discuss the problem of finding sets of stubborn agents that lead to the fastest convergence to equilibrium. By interpreting an opinion that is spread as a message, these "optimal" agents enable its fastest communication. Here as in recent work [**20**] and [**21**], and following earlier work in [**6**], we will formulate an optimization problem based on a random walk on the underlying graph associated with the averaging matrix. The convergence time is asymptotically equivalent to the expected first hitting time of the random walk to the target nodes representing the stubborn agents [**6**],[**3**]. Thus the search for stubborn agents of fixed cardinality can be posed in terms of finding a set for which the first hitting time is smallest.

A search for solutions of these optimization problems involve an exponential number of computations so polynomial time algorithms for approximate solutions are sought. In this situation, some success is possible because the respective objective functions are supermodular (see section 3 for the definition). Supermodularity of the first hitting time was established by Borkar [**6**] and later as a consequence of the work of Clark et al [**11**] and Hunt [**20**]. Recently using the convergence rate as an objective function, supermodularity and convex optimization techniques have been used by Mai and Abed [**26**] to solve the problem we discuss here and its generalizations to multiple competing groups of stubborn nodes. Our strategy is to use the properties of the random walk and take advantage of the graphical structure of optimal and highly optimal sets in order to develop algorithms that can utilize these features. As a consequence, we may eventually identify classes of graphs for which we can efficiently solve or approximate the fast consensus problem and related node selection problems. Indeed although current methods in general deliver approximations that are guaranteed to be within $(1 - 1/e)$ of optimal, provable improvements on that value often rely on assumptions on the curvature and this is difficult to compute in practice. See [**21**],[**26**] for examples where the curvature is used.

A vertex cover of a graph is a subset of vertices that cover all of its edges, i.e. every edge has at least one endpoint in the set. In Section 3, we show that for the uniform random walk (with no self loops) a vertex cover is an optimal set for its cardinality and the same is true for any superset of it. This also holds in general for random walks with no self loops and for lazy versions of these walks (see Section 2 for the definition of lazy). Since a vertex cover (not necessarily minimal) is easy to compute it suffices to restrict the problem to finding an approximately optimal set of cardinality smaller than that of some fixed vertex cover. Furthermore given a vertex cover, all sets of smaller cardinality can be ranked relative to it: the smaller the objective function, the higher the rank. We present a polynomial time generalization of the classic greedy algorithm that begins with a starter set of optimal and nearly optimal sets rather than the best singleton set. The precise definition of a nearly optimal set is given in section 4. For a graph with N vertices and a starter set consisting of singletons, the computational effort is comparable to the greedy algorithm, $O(N * E)$ where E is the computational effort to evaluate the objective function on a subset of V. For a starter set of m element sets it is $O(N^m * E)$. In [**20**] our solution and the greedy algorithm solution were compared using a submodular rank function. The supermodularity and non-increasing properties of our objective function were used to show that the rank of our solution is at least $(1 - \frac{1}{e})$ times the rank of the optimal set and is at least as good as the greedy solution. Moreover our computations show that our solution is often better and sufficient conditions that imply an improved lower bound on the rank can be found in [**21**].

The convergence rate to consensus can also be characterized in terms of the leading eigenvalues of the matrix defining the dynamics. In the discrete time formulation it is the leading eigenvalue of the substochastic matrix corresponding to the non-stubborn nodes (see the work of Borkar [**6**], Como and Fagnani [**14**], Ghaderi and Srikant [**18**]). In the continuous time case, the convergence rate is controlled by the smallest real eigenvalue of the submatrix of the Laplacian restricted to the non-stubborn nodes, i.e. the grounded Laplacian ([**31**]). Upper and lower bounds on this eigenvalue were obtained by Pirani and Sundaram when the graph is undirected [**32**]. Recently Xia and Cao [**40**] established many similar lower and upper bounds for the directed case. The bounds on the eigenvalues obtained by these authors give insight into the graph topology of stubborn nodes. In this paper we will discuss analogous results on our expected hitting time objective function that connect the consensus convergence rate to the graph topology of the non-stubborn nodes. Useful information about the dependence of the rate of convergence on the interaction of graph topology and properties of the random walk follows from recent work of Basu, Hermon and Peres in the mixing theory of reversible Markov chains [**5**] (see section 5.1). We use a result of theirs (Theorem 5.1) to derive upper bounds on the sum of the first hitting times to an arbitrary subset A. No assumptions are made about the number of components in the complement of A, $A^c = V \setminus A$. The bound involves the bottleneck ratio of the boundary of A^c, the number of uncovered edges, i.e. the edges with both endpoints in A^c, and lastly an accessibility measure; a weighted version of centrality measure that arises in a model of binary opinion dynamics in the presence of stubborn agents [**41**]. The number of uncovered edges is a measure of the size of the components and the connectivity of A^c. This interaction of local and global topological properties of the graph with properties of random walks in the context of rate of convergence to consensus is new. To illustrate the usefulness of the upper bound, a simplified version of it is used in section 5 to screen for highly optimal sets without computing the objective function.

2. Notation

Since our paper concerns random walks on a graph we introduce the notation and concepts to be used. An exposition of Markov chain theory can be found in [**24**], while random walks on graphs are discussed in [**3**], [**25**]. A graph is denoted by $\mathcal{G} = (V, \mathcal{E})$, with V the set of nodes and $\mathcal{E}$, its edges. An individual node is indicated by small Roman letters e.g. i, j. The graph is assumed to be strongly connected i.e. for any two nodes there exists a path in the graph that connects them. Throughout our discussion a subset of nodes $A \subset V$, has a complementary set $A^c = V \setminus A$. A node in A^c is called a boundary node if it is adjacent to a node in A. The notation $\mathcal{T}_i$ will be used to denote the connected component of A^c that contains the node i.

The position at any time of a random walker traversing the nodes of a graph in discrete time steps can be described by a random process known as a Markov chain. Here the states of the Markov chain are nodes of the graph and the state at any time is the node the random walker occupies at that time. Given an initial probability distribution on the nodes, the distribution after a single time step can be found in terms of the transition probability matrix P. Using a vector p, let the ith component p_i denote the probability of being at node i at a fixed time step. The position at the next step is described by the probability distribution $p^T P$. The (i,j)-th entry of P , $P(i,j)$ is the conditional probability that random walker steps to j given that the walker was at i in the previous time step. The symbol $\mathbb{P}$ is used to denote the probability distribution associated with the paths of the Markov chain. The expected value or average of a random variable X with respect to $\mathbb{P}$ is $\mathbb{E}[X]$.

Notions of conditional expectation are also used. The notation for the expected or mean value of a random variable X given the random walker is initially at node i is $\mathbb{E}_i[X]$. More generally if the initial position of the walker is defined by a probability distribution μ, then the expected value of X given μ is $\mathbb{E}_\mu[X]$. When the Markov chain is irreducible and aperiodic, the distribution of possible positions of the random walker after a long time is described by the so-called stationary distribution, denoted by the row vector π. In addition we will assume the Markov chain is reversible so that $\pi(i)P(i,j)=\pi(j)P(j,i)$. The commonest example of such a process is the simple uniform random walk whose transition probabilities are:

$$P(i,j) = \begin{cases} \frac{1}{deg(i)}, & \text{if } j \sim i; \\ 0 & \text{otherwise.} \end{cases} \tag{2.1}$$

A node i is adjacent to j iff (i,j) or $(j,i) \in \mathcal{E}$. The notation for this relation is $i \sim j$. The results of this paper apply to any walk on a directed graph with no self-loops that is described by a reversible Markov chain. Note that we are allowing weighted edges i.e. a non-symmetric adjacency matrix. The results on first hitting times and optimality can be extended to lazy versions of the random walk as well because the mean values of these random variables for the non-lazy case can be obtained by dividing by 2. A lazy version of the random walk governed by the transition matrix P has a transition matrix $(I+P)/2$. The same nodes and edges are used as in the original Markov chain and the stationary vector is the same for both chains. In this work $P(i,j)$ and $P(j,i)$ are assumed to be positive if (i,j) is an edge and $i \neq j$. All the computations described in this paper used the uniform random walk or its lazy version.

In section 3 we will discuss the random variable that describes the first time a random walker visits a subset A, i.e. the first hitting time of A, $T_A \triangleq \min\{t \geq 0 : X_t \in A\}$, where X_t is the position of the walker at time t. The dynamics of the random walk before time T_A is described by the principal submatrix of P obtained by crossing out the rows and columns of P corresponding to the nodes of A. We denote this matrix by P_A in section 3. To describe the process of the random walk before the set A is reached it is useful to count the number of times an edge e is crossed. We define n_e be the number of times a random walker crosses e before time T_A.

3. Problem setting

Although the rate of convergence to consensus given a subset A of stubborn or leader nodes is determined by the leading eigenvalue of the sub-stochastic matrix P_A, the arguments of Clark et al [**10**] can be modified to show that the convergence rate of the error has an upper bound given by the mean or expected first hitting time to A. For the convenience of the reader a derivation of this connection is given in the Appendix. Thus following [**6**] (equation 8), we seek a set that maximizes the convergence rate by minimizing F, the sum of the mean first hitting times of random walkers outside the set. Thus for the set A let,

$$F(A) = \sum_{i \notin A} \mathbb{E}_i[T_A]. \tag{3.1}$$

If A is an effective target set for the random walk then $F(A)$ will be small and this implies fast convergence to consensus. Thus subject to a constraint on the cardinality K of the sets to be considered, we define the optimization problem,

$$\min_{A \subset V,\, |A| \leq K} F(A). \tag{3.2}$$

Since $\max_i \mathbb{E}_i[T_A] \leq F(A) \leq |V \setminus A| \max_i \mathbb{E}_i[T_A]$, we know that F is the rate of convergence up to a constant (see Appendix). The optimization problem as stated is quite difficult so approximations are sought that exploit properties of F. We observe that if $A \subseteq B$, then $F(A) \geq F(B)$. Thus F is a non-negative real valued non-increasing set function. Secondly it can be proved [**6**],[**10**] that F is supermodular. That is, $A \subseteq B \subseteq V$ and $j \in V$, $F(A) - F(A \cup j) \geq F(B) - F(B \cup j)$. These properties also follow from an explicit formula for $F(A) - F(A \cup j)$ derived in [**21**]. As a consequence $-F$ is submodular and when it is normalized (see section 4.2), the greedy algorithm produces an approximation to the solution of problem (3.2) that is within $(1 - \frac{1}{e})$ of optimal [**28**]. Given a set A, a standard result in Markov chain theory [**24**] can be used to calculate $F(A)$. Let $h(i, A) = \mathbb{E}_i[T_A]$. Then it is the ith component of the vector solution H of ,

$$H = \mathbf{1} + P_A H. \tag{3.3}$$

Here P_A is the matrix that results from removing from P, the rows and columns corresponding to the nodes in A and $\mathbf{1}$ is the column vector of $N - |A|$ ones. The value $F(A)$ is then computed by summing the components of H. From equation (3.3) it immediately follows that if the Markov chain has no self loops any vertex cover A is an optimal set, i.e. a solution of problem (3.2) for $K = |A|$. To see this recall that A is a vertex cover if every edge of $\mathcal{G}$ has at least one endpoint in A. This implies that for every $i \notin A$, $h(i, A) = 1$. However equation (3.3) implies that $h(i, A) \geq 1$, hence $F(A)$ is minimal. Since a superset of a vertex cover is also a vertex cover we see that if a vertex cover of cardinality C is available, then the problem is solved for $K \geq C$ and only needs to be solved for $K < C$. A vertex cover (not necessarily minimal) can be easily computed. In particular, the vertices of a maximal match are a vertex cover and a straightforward algorithm of complexity $O(|\mathcal{E}|)$ produces a set no larger than twice the cardinality of the minimum vertex cover [**13**].

4. Optimal and near optimal sets

There are two observations that are key to our approach to problem (3.2). The first as discussed in Section 3 is that vertex covers are optimal sets for their cardinality. The second is that a vertex cover often contains subsets that are optimal sets (see however [**20**] for an example of a vertex cover that contains no optimal sets as proper subsets). Thus given a vertex cover of cardinality C and a $K < C$, the search for optimal sets is restricted to near optimal (or optimal) sets of cardinality $\leq K$. We expect that optimal sets of size K much like vertex covers will contain subsets that are optimal or in some sense near optimal. Then generalizing the classic greedy approach, each set in a starter set (rather than the best singleton) is greedily extended in a stepwise fashion until a set of cardinality K is reached. The offered approximation is the set with the smallest value of F among all the extended sets. It remains then to make the term "near optimal" precise. We will rank all non-empty subsets $A \subset V : |A| \leq C$ with a ranking function r, defined as,

$$r(A) = \frac{F_{max} - F(A)}{F_{max} - F_{min}} \tag{4.1}$$

where $F_{max} = \max_{\emptyset \neq A \subseteq V, |A| \leq C} F(A)$, and F_{min} is the corresponding minimum. F_{min} can be calculated by computing the F value of the vertex cover of cardinality C. We define F_{max} to be the maximum value of F among all one element subsets. We can assume that $F_{max} \neq F_{min}$. Indeed if this was not the case, $F(A)$ would have the same value for any non-empty A of cardinality less than C and hence any such set would be a solution. The function r assigns a rank to all subsets A with $|A| \leq C$. For example if A is optimal and $|A| = C$ then $r(A) = 1$, conversely for the worst performing set $r(A) = 0$. We can now define a class of near optimal sets relative to a vertex cover of cardinality C.

Given constants ν, $0 < \nu < 1$ and $C > 0$, the non-empty set

$$L_{\nu,C} \triangleq \{A : A \subseteq V, |A| \leq C,\, r(A) \geq \nu\}, \tag{4.2}$$

defines a class of optimal and near optimal subsets with degree of near optimality depending on ν and C. Putting aside the question of how to find sets that are in $L_{\nu,C}$, let m be the size of the cardinality of the smallest set in $L_{\nu,C}$. Then it contains all the optimal solutions to (3.2) for $m \leq K \leq C$. Thus, in this range, the search for optimal sets can be restricted to $L_{\nu,C}$. If a graph has a tree like or spoke and wheel structure, this can mean a substantial reduction in search effort. Moreover since F is a non-increasing set function, when the set A is in $L_{\nu,C}$ then for any element $a \in V$, $A \cup \{a\} \in L_{\nu,C}$. Thus any set of cardinality K that is obtained by successive one element additions to a set of cardinality m in $L_{\nu,C}$ is also in $L_{\nu,C}$. In particular, a set that is the result of the application of the greedy algorithm, has a rank that is at least ν. In the definition of $L_{\nu,C}$ we can assume there is a fixed vertex cover so that C is known. If a degree of optimality ν is prescribed, it dictates the value of m. So for example if $\nu > \max_{|A| \leq K|} r(A)$, then we must have $m > K$. Conversely m will constrain ν to values that are achievable by sets of cardinality m. That is,

$$\nu \leq \max_{|A| \leq m} r(A). \tag{4.3}$$

To find sets of some predetermined cardinality m that are in $L_{\nu,C}$ we will start by evaluating F on sets of some predetermined size m. The value of ν must therefore satisfy condition 4.3. In practice because of the computational effort, we choose $m \leq 3$. Our strategy for solving problem 3.2 is then to create a class of starter sets $\mathbf{S} \subseteq \{X \subset L_{\nu,C}\,, |X| = m\}$ for some ν in the constrained range.

If S^* is the offered solution resulting from our method it was shown in [**20**],[**21**] that it is at least as good as the greedy solution and the degree of improvement can be calculated from S^* itself. When more is known calculation of S^* is not needed to estimate the degree of improvement. Suppose S with $|S| \leq K$ is an extension of a set in $\mathbf{S}$, and the normalized rank $\rho(S)$ (see section 4.2) satisfies $\rho(S) > \eta \geq \rho(S_g)$. Then the degree of improvement has an explicit lower bound in

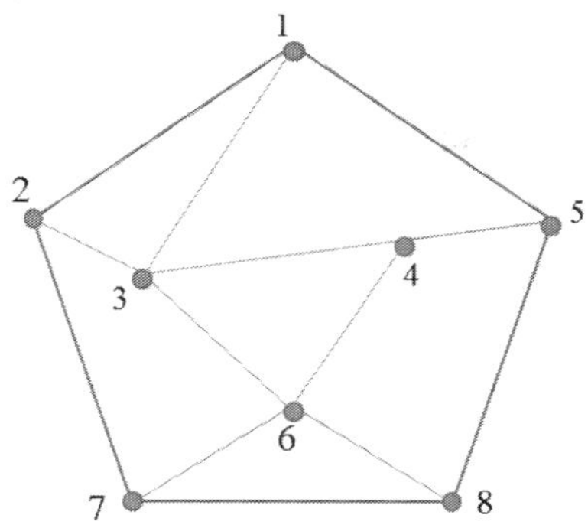

FIGURE 1. Graph with $N = 8$ vertices and $E = 13$ edges.There are vertex covers with 5 elements.

TABLE 1. Greedy extension of starting set $\mathbf{S}$ ($m = 2$) for graph in Figure 1 The optimal sets are in boldface.

K=2	3	4	5
1, 6	1,6,8	1,6,8,3	1,6,8,3,4
3, 8	3,8,1	3,8,1,6	3,8,1,6,4
3,7	**3, 5, 7**	**3, 5, 6, 7**	**1, 3, 5, 6, 7**
2,6	**2, 6, 5**	**2, 6, 5, 3**	**2, 6, 5, 3, 7**
1,8	1,8,3	1,8,3,6	1,8,3,6,4
4,8	4 8 2	4 8 2 1	**4, 8, 2, 1, 6**

terms of η. Convenient candidates for such an S are subsets of the reference vertex cover.

Note that when $m = 1$, $\mathbf{S}$ is a collection of singletons, the maximum possible value of ν selects the best singleton as the starter set. Thus our method contains the classic greedy algorithm as a special case.

4.1. Algorithm description and examples. To solve the problem (3.2) for $K < C$, we choose a collection of sets $\mathbf{S} \subset L_{\nu,C}$. Each set in $\mathbf{S}$ has cardinality m, the minimum cardinality of sets in $L_{\nu,C}$. Note that the exact solution is an element of this class whenever $m \leq K \leq C$. The output of this method, i.e. the offered approximation is the best set that results from a greedy extension of each set in $\mathbf{S}$, to a set of cardinality K. Since exact calculation of F occurs for sets up to cardinality m only, the method has complexity $O(N^m * E)$ where E is the complexity of evaluating F for a single subset. Thus when $m = 1$ the complexity of our method and the classic greedy algorithm are the same but is greater when $m > 1$. For example in [**21**] we assumed that the value of F was obtained by solving equation (3.3) by matrix inversion of an $N \times N$ matrix. In this case $E = O(N^3)$. Thus this method is very expensive for graphs with $N = O(10^3)$ vertices without some modification.

The spectral sparsification theory and associated development of nearly linear time algorithms for solving certain classes of systems of equations with Laplace matrices in the work of D.A. Spiegelman and S.H. Teng [**36**], opens the possibility of a significantly reduced complexity for evaluating F at a single set namely $E = O(N \log^2 N)$. Thus the complexity of our method is then $O(N^2 \log^2 N)$ and

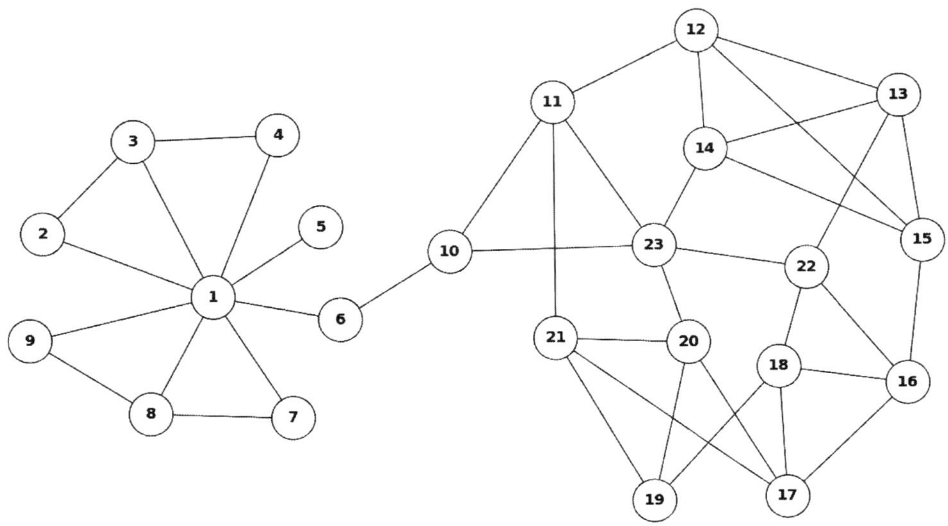

FIGURE 2. Graph with $N = 23$ vertices, $E = 97$ edges. A 12 element vertex cover is $\{1, 3, 8, 10, 12, 13, 15, 16, 18, 20, 21, 23\}$

$O(N^3 \log^2 N)$ for $m = 1$ and $m = 2$ respectively. The relation between the linear equation (3.3) used to calculate F and the systems treated by fast Laplacian solvers can be easily seen. Recall that $\mathbf{L}$, the normalized Laplacian discussed by Spiegelman and Teng is,

$$\mathbf{L} = D^{-1/2} \mathbb{L} D^{-1/2} \tag{4.4}$$

where $\mathbb{L} = D - \mathbb{A}$, D is an $N \times N$ diagonal matrix whose ith diagonal entry is d_i=deg(i), and $\mathbb{A}$ is the adjacency matrix of the graph. For the uniform random walk $I - P = D^{-1}\mathbb{L}$.

In section 5.3, we discuss a heuristic method of selecting highly optimal sets of fixed cardinality without calculating F and therefore without solving a linear equation. This approach can be used to create a starter set at a much lower computational cost.

We illustrate the method for solving problem (3.2) using the graph shown in Figure 1. Here $N = 8$ and $E = 13$. Although the graph has several vertex covers of cardinality 5, it was easy to use equation (4.1) to compute $r(A)$ for $L_{1/2,7}$. Here $m = 2$ and to implement the method we used a starter set $\mathbf{S}$ consisting of all 24 of these cardinality 2 sets. Ordering the elements of $\mathbf{S}$ in terms of the corresponding values of F, Table 1 shows the results of the greedy extension of the sets that occur in positions $i = 1, 2, 3, 4, 8, 15$, for $K = 2$ through 5. The results of sets in positions $i = 5, 6$ and 9 through 14 are omitted because because they produce the same optimal sets as $i = 3, 4$.Others are omitted because none of their extensions in this K range are optimal.

The next example (see Figure 2) appears in a study of node centrality [**17**]. It has 23 nodes graph and there is a vertex cover with 12 elements (see text) . Here we will show how our method generates offered solutions to the problem (3.2) for $1 \leq K \leq 12$. The minimum cardinality of sets in the class of near optimal sets $L_{.50,12}$ is $m = 1$ so determining the exact solutions of equation (3.3) is only needed

TABLE 2. Greedy extension of near optimal singletons of starting set for graph in Figure 2.The optimal sets are in boldface.

K=1	2	3	4	5	6
23	**23, 1**	**23, 1, 16**	23,1,16,21	**23, 1, 16, 21, 12**	
17	17,1	17,1,23	**17, 1, 23, 13**		
13	13,1	13, 1, 20			
18	18, 1	18,1,23	18,1,23,12	18,1,23,12,	**18, 1, 23, 12, 15**

K=7	8	9
		23, 1, 16, 21, 12, 18, 3, 8, 13
18, 1, 23, 12, 21, 15, 3	**18, 1, 23, 12, 21, 15, 3, 8**	

for $K = 1$. Using the 15 one element sets of $L_{.50,12}$ as a starter set, all of the optimal sets up to $K = 12$ are obtained by their greedy extension. The optimal sets for $K = 10, 11, 12$ are greedy extensions of the optimal set for $K = 9$. Moreover as shown in Table 2 , all of the optimal sets(shown in boldface) can be obtained from the extension of just 4 elements in the starter set. Entries for non-optimal sets for K larger than 6 are not shown. Note that as in our first example, the results of the classic greedy algorithm failed to identify some of the optimal sets.

Our final example is a model of a Internet Service Provider (ISP) network [**15**]. It has 218 nodes and there is a vertex cover with 41 nodes (see Figure 3). The optimal sets for $2 \leq K \leq 41$ were approximated using the algorithm and Figure 4 shows the results of using two different starter sets. For the near optimal class $L_{7/8,41}$, we have $m = 1$, and for $m = 2$, a second starter set comes from $L_{.93,41}$. The performance of each starter set is evaluated by calculating the ratio $\frac{F_{greed}(K)}{F_{alg}(K)}$ for each value of K. Here $F_{greed}(K)$ is the estimate of the optimal value of F for cardinality K obtained by the classical greedy method and $F_{alg}(K)$ is the estimate of the optimal value using the algorithm. As a consequence of Corrollary 1 in [**21**], this ratio is always at least 1. The generally larger values of the ratio for $L_{.93,41}$ show the larger improvement over the greedy solution in comparison to $L_{7/8,41}$.

4.2. Quality of the approximation. Following Ilev ([**22**]) when F is supermodular, the value for the the empty set can be defined as

$$0 \leq F(\emptyset) = \max_{X \cap Y = \emptyset, X, Y \subseteq V} F(X) + F(Y) - F(X \cup Y) < \infty \tag{4.5}$$

Thus by the definition of r , $r(\emptyset) = \frac{F_{max} - F(\emptyset)}{F_{max} - F_{min}}$. This means the function ρ defined on sets A by $\rho(A) = r(A) - r(\emptyset)$ is bounded, submodular and non-decreasing and satisfies $\rho(\emptyset) = 0$. Also note that since $F(\emptyset) \geq F_{max}$, it follows that $\rho(A) \geq 0$ for all $A \subseteq V$. Following the nomenclature of [**39**], ρ is called a normalized rank function. Since ρ is an affine function of F, the optimization problem (3.2), is equivalent to the problem of finding the set that maximizes the normalized rank subject to the same constraints. This is a special case of the general problem first considered by Nemhauser, Wolsey and Fisher. Using their result ([**28**], Section 4) we showed that the ratio of the normalized ranks of the classic greedy solution and the optimal solution has a lower bound of $(1 - \frac{1}{e})$ [**21**]. Let us now assume that the m stage greedy solution is in $\mathbf{S}$. This is not a serious restriction for if it is not there it can always be added.

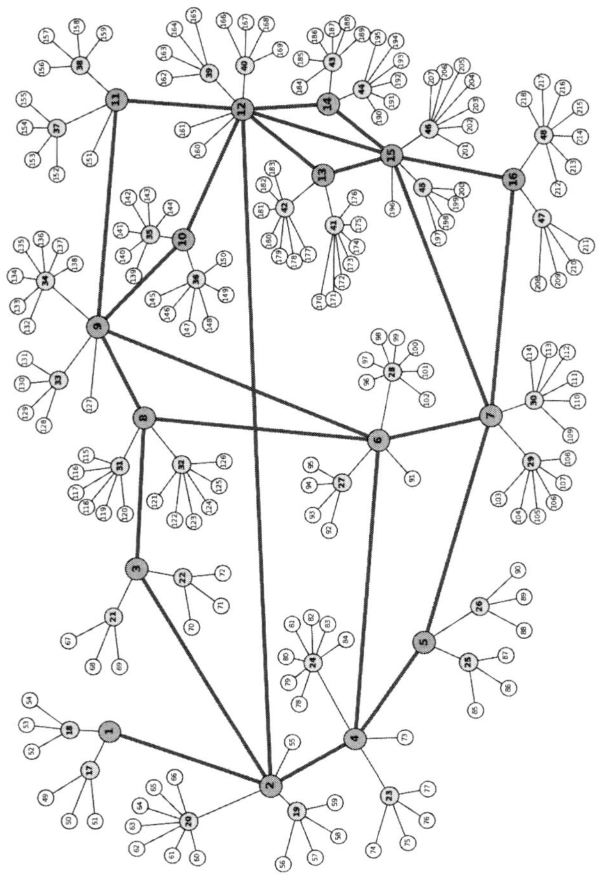

FIGURE 3. A compressed version of the Abilene Network: N=218 vertices, E=226 edges,vertex cover is $\{2, 3, 4, 6, 7, 9, 12, 15, 17...48\}$

PROPOSITION 1. *If S^* is the offered solution constructed by the method described in sections 4, 4.1, then*

$$\rho(S^*) \geq \left(1 - \frac{1}{e}\right) \rho\left(\mathcal{O}_K^*\right) \tag{4.6}$$

where $\mathcal{O}_K^*$ is an optimal set of cardinality K.

The proof uses the fact that the greedy solution S_g of cardinality K satisfies the inequality $\rho(S_g) \geq \left(1 - \frac{1}{e}\right) \rho(\mathcal{O}_K^*)$.

A $(1 - \frac{1}{e})$ lower bound similar to Proposition 1 was established by Borkar et al in [**6**]. Specifically it is a lower bound on the ratio of $F(S_g(a)) - F(\{a\})$ to $F(\mathcal{O}_k^*) - F(\{a\})$, where $S_g(a)$ is the result of the greedy algorithm starting with singleton a.

5. An upper bound and a surrogate for F

In this section we derive upper bounds for $F(A)$ in terms of some topological aspects of graphs induced by A and $V \setminus A$. As a byproduct we also derive a surrogate

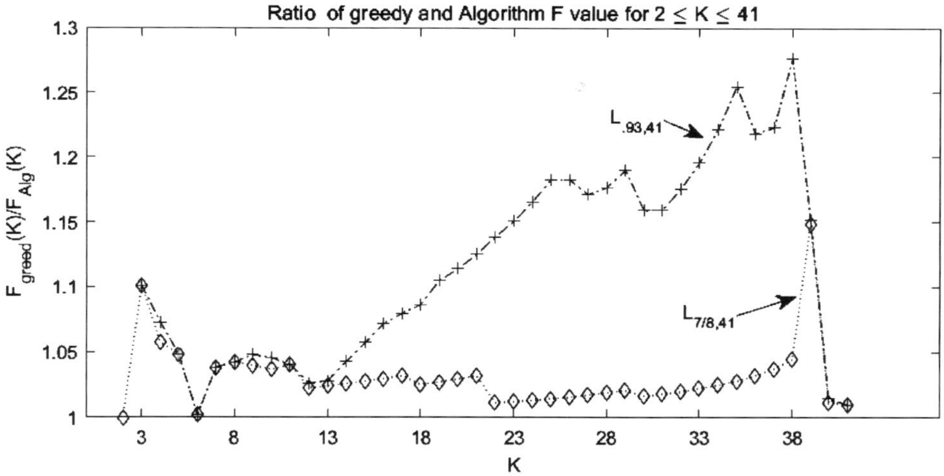

FIGURE 4. Comparison of F values calculated by Algorithm with Greedy solution values for starter sets $L_{7/8,41}$ and $L_{.93,41}$ for the graph shown in Figure 3

set function from the general upper bound of section 5.1 that can be used to select desirable candidates for the solution of problem (3.2) without directly computing F. We first derive a representation of $\mathbb{E}_i[T_A]$ in terms of the number of distinct edges visited by the random walker encountering A. Starting at $i \notin A$ the formula will involve the expected number of times a directed edge of the form $e = (j,k)$, $j,\ k \notin A$ is visited plus the expected number of times an edge of the form $f = (j,k)$, $j \notin A, k \in A$ is visited. Since such an edge ending in A can only be visited once, and since this occurs with probability one, the expected number of times such an edge is visited is 1.

Therefore let n_e be the number of times a random walker starting at $i \notin A$ crosses an edge e before T_A. For a non-lazy random walk with no self-loops (for example the uniform random walk on $\mathcal{G}$), the elapsed time between the start of a walk at $i \notin A$. and the time of first arrival in A is the sum of unit time steps where the walker either moves along an edge joining A^c to A and is absorbed (the expected number of such steps being 1) or moves along an edge e that has both endpoints in A^c. Such an edge is called uncovered. Thus,

$$T_A = 1 + \sum_{n=0}^{\infty} \{ \sum_{e\ uncovered} \mathbf{1}_e(X_n, X_{n+1}) \mathbf{1}(n < T_A) \}. \tag{5.1}$$

Thus the expected value of T_A, starting at i can be written as,

$$\mathbb{E}_i[T_A] = 1 + \mathbb{E}_i[\sum_{n=0}^{T_A - 1} (\sum_{e\ uncovered} \mathbf{1}_e(X_n, X_{n+1}))]. \tag{5.2}$$

Now equation (5.2) can be rewritten as,

$$\mathbb{E}_i[T_A] = 1 + \mathbb{E}_i[\sum_{e: uncovered} n_e]. \tag{5.3}$$

REMARK. Uncovered edges can be divided into edges comprising the connected components of $V \setminus A$. So for a fixed node i, equation (5.3) can be rewritten as

$$\mathbb{E}_i[T_A] = 1 + \mathbb{E}_i[\sum_{e: e \in \mathcal{T}_i} n_e], \tag{5.4}$$

where $\mathcal{T}_i$ is the component containing i.

5.1. Upperbound on $F(A)$. To obtain an upper bound on F for a general set, we will use a recent result of Basu et al ([**5**]) from the mixing theory of Markov chains. We assume the chain is irreducible and reversible therefore there exists a unique stationary distribution for the chain, π. Basu et al discovered a formula for the expected first hitting time to a set A starting from an initial distribution of points in A^c on the boundary of A that were in A at the previous time step. The expected first hitting time is equal to the bottle neck ratio (sometimes called the conductance). It is used to define Cheeger's constant, a lower bound on the mixing time of the chain ([**25**]).

The bottle neck ratio [**25**] of set B, $\Phi(B)$ is defined as,

$$\Phi(B) = \frac{\sum_{x \in B,\, y \in B^c} \pi(x) P(x, y)}{\sum_{x \in B} \pi(x)} \tag{5.5}$$

For the special case of the uniform random walk this is,

$$\Phi(B) = \frac{|\partial B|}{\sum_{b \in B} deg(b)} \tag{5.6}$$

Here $|\partial B|$ is the number of edges connecting $B = A^c$ to A.

Basu et al obtained an expression for the expected first hitting time to a set $A \subset \Omega$, for a random walk starting at a node in A^c adjacent to A. Specifically consider the possible destinations in $B = A^c$ of a walk that at the previous time step was in A. They define a probability distribution on the possible landing points in A^c, $\Psi_{A^c}(y) = P_{\pi_A}[X_1 = y | X_1 \in A^c]$. Here π_A is the stationary probability distribution restricted to A.

$$\pi_A(x) = \frac{\pi(x) \mathbf{1}_A(x)}{\pi(A)}$$

For an irreducible, reversible chain the expected first hitting time to A given the initial distribution Ψ_{A^c}, is expressed in terms of the bottleneck ratio.

THEOREM 5.1. [**5**] *Let (Ω, $\mathbb{P}$, π) be a finite irreducible reversible Markov chain. Let $A \subsetneq V$ with complement B be given. Then,*

$$\mathbb{E}_{\Psi_B}[T_A] = \frac{1}{\Phi(B)}. \tag{5.7}$$

Suppose a random walk begins at b in the support of Ψ_B, and then visits a node $i \notin A$ before reaching A for the first time. By the Strong Markov property, the number of times an uncovered edge e is crossed by the walk after the first visit to i but before T_A, is the number of times e is crossed when starting at i before T_A. Let T_i^y for $y \notin A$ be the time for a random walker starting initially at y, to arrive at i for the first time. If i is a non-isolated point of A^c, and $y \in \mathcal{T}_i$, then such a path exists with positive probability. The hypotheses on the Markov chain are the same as those in Theorem 5.1.

LEMMA 5.2. *Let $i \in B = A^c$, be a non-isolated point in B and suppose y is adjacent to A. Then*

- $y \in supp(\Psi_B)$
- *If $y \in \mathcal{T}_i$ (the connected component of $V \setminus A$ that contains i) then $T_i^y < T_A^y$ with positive probability*

PROOF. Since y is adjacent to A there is an edge (a, y) with $a \in A$, thus $p(a, y) > 0$. Using the definition of Ψ_B, one can write,

$$\Psi_B(y) = \frac{\sum_{a \in A} \pi(a)p(a, y)}{\sum_{b \in B,\, a \in A} \pi(a)p(a, b)} \tag{5.8}$$

Thus, $\Psi_B(y) > 0$ since $\pi(a) > 0$ for all a. To show that $T_i^y < T_A^y$ with positive probability, note that when $i = y$, then the inequality holds with probability 1. If $i \neq y$, then the connectedness of $\mathcal{T}_i$ and the reversibility of the Markov chain imply the existence of a path from y to i that lies entirely in A^c. Thus there is a sample path starting at y and ending at i and since the transition probability on any edge of the path is positive, the sample path has positive probability by the Chapman Kolmogorov equality. For such a path $T_A^y = T_i^y + T_A^i$ so the claimed inequality holds with positive probability. □

REMARK. Let δA be the nodes in A^c that are adjacent to A. The lemma implies that for fixed non-isolated $i \notin A$, any $b \in \delta A \cap \mathcal{T}_i$ is in the support of Ψ_B. If i is isolated then it is automatically in $supp(\Psi_B)$ and we have $y = i$ that is, $T_i^y = T_i^i < T_A^y$ with probability one.

We now turn to the derivation of an upper bound for $\mathbb{E}_i[n_e]$. The upper bound accounts for the likelihood that $i \notin A$ is visited by a random walk started at b before the walk hits A for the first time.

PROPOSITION 2.

$$\mathbb{E}_i[n_e] \leq \frac{\mathbb{E}_{\Psi_B}[n_e]}{\sum_{b \in \delta A} \Psi_B(b)\mathbb{P}(T_i^b < T_A^b)} \tag{5.9}$$

PROOF. Let e be an uncovered edge in the component $\mathcal{T}_i$. We define n_e^i to be the number of times a random walk path crosses an uncovered edge e during the time period, $T_i \leq t \leq T_A$ with $n_e^i = 0$ if $T_i > T_A$. It can be seen that, for $b \in supp(\Psi_B)$,

$$\begin{aligned} \mathbb{E}_b[n_e] \geq \mathbb{E}_b[n_e^i] = \mathbb{E}_b[n_e^i; T_i^b < T_A^b] = \\ \mathbb{E}_b[n_e^i \,|T_i^b < T_A^b] \cdot \mathbb{P}[T_i^b < T_A^b]. \end{aligned} \tag{5.10}$$

By the Strong Markov property, $\mathbb{E}_b[n_e^i \,|T_i^b < T_A^b] = \mathbb{E}_i[n_e]$. Thus

$$\mathbb{E}_b[n_e] \geq \mathbb{E}_i[n_e] \cdot \mathbb{P}[T_i^b < T_A^b]. \tag{5.11}$$

Multiplying each side of inequality (5.11) by $\Psi_B(b)$ and adding over all $b \in supp(\Psi_B)$ results in the inequality

$$\mathbb{E}_{\Psi_B}[n_e] \geq \left(\sum_{b \in supp(\Psi_B)} \Psi_B(b)\mathbb{P}[T_i^b < T_A^b] \right) \mathbb{E}_i[n_e] \tag{5.12}$$

By Lemma 5.2 and the remark that follows its proof, the sum on the right hand side of (5.12) is non-zero and the set $supp(\Psi_B)$ can be replaced by δA. Thus the statement of the proposition follows from (5.12). □

An upper bound on $\mathbb{E}_i[T_A]$ can be deduced from Proposition 2. First we have for any uncovered edge e, $\mathbb{E}_{\Psi_B}[n_e] \leq \mathbb{E}_{\Psi_B}[T_A]$. Thus from equation (5.3) it follows that,

$$\mathbb{E}_i[T_A] \leq 1 + \mathbb{E}_{\Psi_B}[T_A] \left(\sum_{e \in \mathcal{T}_i} p_e \right) \left(\frac{1}{d(i, A)} \right) \tag{5.13}$$

where $p_e = 1$ if e is an uncovered edge and is 0 if e is covered, and $d(i, A) = \sum_{b \in \delta A} \Psi_B(b) \mathbb{P}(T_i^b < T_A^b)$.

The upper bound on $F(A)$ then follows from inequality (5.13).

COROLLARY 1. *For $B = A^c$,*

$$F(A) \leq N - |A| + \frac{1}{\Phi(B)} \left(\sum_{e \in \mathcal{E}} p_e \right) \left(\sum_{i \notin A} \frac{1}{d(i, A)} \right). \tag{5.14}$$

REMARK. It is interesting to note that nothing about the connectivity $V \setminus A$ was assumed in the derivation of the upper bound in (5.14). Equality holds when A is a vertex cover since in this case $\sum_{e \in \mathcal{E}} p(e) = 0$.

The inequality (5.14) has two useful consequences. First, it gives insight into the conditions for fast consensus. Like the work of Pirani ([**31**]), Pirani and Sunderam ([**32**]), Xia and Cao ([**40**]) on bounds for the smallest eigenvalue of grounded Laplacians, the bound depends on graph theoretic properties of the (leader) stubborn nodes and the connectivity of the residual graph of (follower) non-stubborn nodes. Secondly in the uniform random walk case there is a simplified form of (5.14) that is an effective tool for identifying highly optimal sets without actual computation of F (section 5.3). When all the candidate sets have the same cardinality, it can be used to select nearly optimal K element sets. This is a natural way to improve the offered approximation of the greedy extension algorithm or as we mentioned earlier a computationally cheaper way to generate a starter set.

The upper bound in Corollary 1 depends on three somewhat independent graph properties of A and $V \setminus A$. The bottle neck ratio (a consequence of Theorem 1) counts the fraction of edges joining A and $V \setminus A$, and the second the number of uncovered edges, depends of the ability of A to minimize the connectivity of the residual set $V \setminus A$. In this respect, an optimal or near optimal A resembles a near optimal solution of the partial vector problem ([**7**]) or critical node problem ([**4**]); important discrete optimization problems with applications in network security and communication. The third factor is node specific. The quantity $d(i, A) = \sum_{b \in \delta A} \Psi_B(b) \mathbb{P}[T_i^b < T_A^b]$, measures the accessibility of the node $i \notin A$, to A. It is the weighted probability that i can be reached by a random walker starting at a node b that is adjacent to A before the walker visits A for the first time. Nodes i that are remote have a small $d(i, A)$ value resulting in a large contribution to the bound.

The accessibility quantity is very closely related to a centrality parameter that arises in the study of opinion dynamics in a social network with stubborn agents [**41**]. The context is a binary voter model with rival sets of stubborn agents holding different opinions (labeled 0 and 1 respectively). The authors define the quantity $r(i,A) = \sum_{j \notin V_0 \cup V_1} \mathbb{P}[T_i^j < T_A^j]$, where $A = V_0 \cup V_1$ and V_0, V_1 are the nodes holding opinions 0 and 1 respectively. Thus, $r(i,A)$ measures how influential i is in shaping the opinions of other non-stubborn nodes in the presence of stubborn agents in V_0 and V_1 (page 19.13, page 19.9 [**41**]). In the presence of a single fixed set of stubborn agents holding opinion 0, $r(i,A)$ is an unweighted version of $d(i,A)$. The authors consider the best location for a rival stubborn node i holding opinion 1. This reduces the problem to maximizing an objective function (the probability of absorption by i before absorption by the set A). The optimal placement problem is treated in a manner similar to what we do here. However we do not use a gluing argument as in [**41**], thus we can analyze the case where $V \setminus A$ has multiple components. This can be useful when the number of connected components of $V \setminus A$ is known for example when starter sets are good approximations of the solution of the critical node problem [**4**].

There are connections between the $d(i,A)$ and the eigenvector components of the zero eigenvalue of the graph Laplacian [**18**], [**31**], [**40**]. This connection merits discussion because the eigenvalue approach may lead to methods of altering the graph topology to facilliate convergence and other desirable network behavior. In the next section, the theory of absorbing Markov chains is used to explain the link between $\{d(i,A) : i \in A^c\}$ and the eigenvectors of the graph Laplacian matrix. The use of absorbing Markov chains in consensus models is not new. Pirani used the theory to derive upper and lower bounds on the maximum eigenvalue of the inverse of the grounded Laplacian associated with A. Researchers (see e.g. [**1**], [**18**], [**31**],[**30**], [**14**], [**41**]) have also used the theory to derive representations for the consensus value itself.

5.2. Connections with the graph Laplacian. Let P be the transition matrix of a Markov chain with n absorbing states corresponding to nodes $\mathcal{S} = 1 : n$. The graph of the Markov chain is a directed graph with no outgoing edges at nodes $\mathcal{S}$. The Laplacian matrix for the graph is $\mathcal{L} = I - P$. After appropriate labeling of nodes P can be put in the standard form for absorbing chains ([**24**])

$$P = \begin{pmatrix} I & O \\ R & Q \end{pmatrix} \tag{5.15}$$

where $n = |\mathcal{S}|$, I is an $n \times n$ identity matrix, O is a $n \times (N-n)$ matrix of zeros, R is an $(N-n) \times n$ matrix and Q is an $(N-n) \times (N-n)$ matrix. The vector $x \in \mathcal{R}^N$ is an eigenvector of $\mathcal{L}$ corresponding to the eigenvalue $\lambda_{\mathcal{L}} = 0$ if and only if it is an eigenvector of P belonging to $\lambda_P = 1$. The algebraic multiplicity of $\lambda_{\mathcal{L}}$ is n and it is semisimple ([**2**],[**9**]). Therefore the eigenspace is spanned by n independent vector solutions of $\mathcal{L}x = \mathbf{0}$. We now turn to a discussion of how these components are very closely related to the upper bound of F in Corollary 1. Using equation (5.15), and dividing the vector x into components corresponding to absorbing and non-absorbing states $x = \begin{pmatrix} \hat{x} \\ \nu \end{pmatrix}$ one obtains,

$$(I - Q)\nu = R\hat{x}. \tag{5.16}$$

Here I is an $(N-n) \times (N-n)$ matrix, $\nu \in \mathcal{R}^{N-n}$ and $\hat{x} \in \mathcal{R}^n$. The eigenspace of $\lambda_{\mathcal{L}} = 0$ is spanned by vectors $x_i = \begin{pmatrix} \hat{e}_i \\ \nu_i \end{pmatrix}$, $i = 1, \cdots n$. For each i the vector $\hat{e}_i$ is

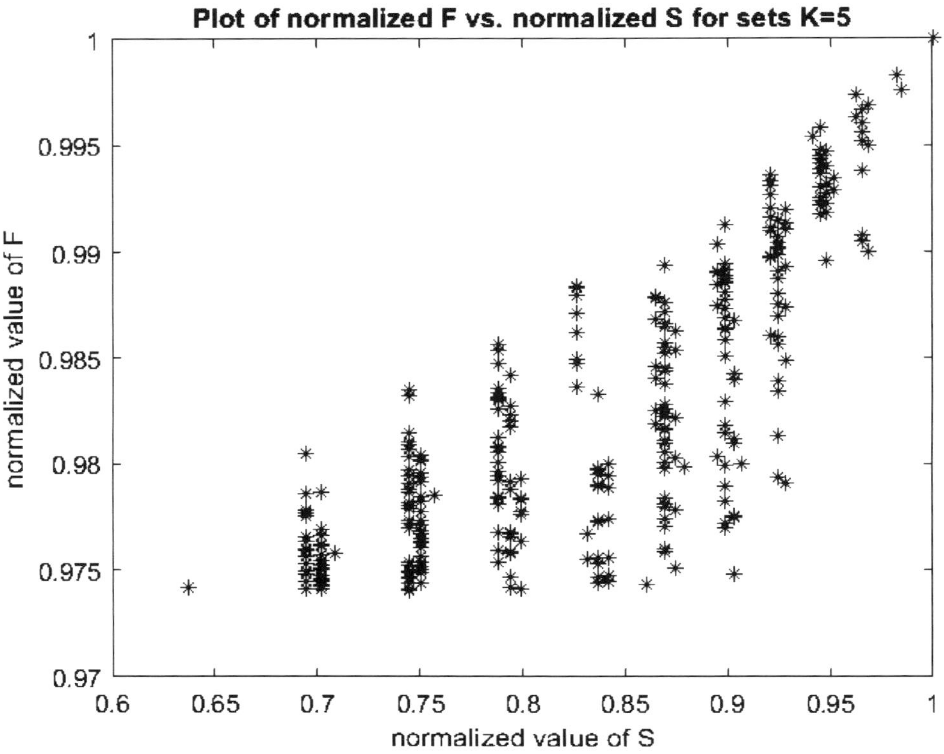

FIGURE 5. Plot of normalized values of F versus normalized surrogate value S. Values for top 8000 (of 33,649) K=5 sets of graph in Figure 2 are plotted every 25 horizontal steps for readibility

the vector with component 1 in the ith position and 0 elsewhere. It represents the absorbing node i and ν_i is the solution of (5.16) when $\hat{x} = \hat{e}_i$. Since x_i is also an eigenvector of P for eigenvalue 1, the matrix $\Lambda = \begin{bmatrix} x_1 & \cdots & x_n \end{bmatrix}$ satisfies $P\Lambda = \Lambda$. The submatrix consisting of the last $N - n$ rows of Λ is $H = \begin{bmatrix} \nu_1 & \cdots & \nu_n \end{bmatrix}$. This matrix has a probabilistic interpretation that is relevant to Corollary 1. Let s be a state in $\mathcal{S}$ and let $b \notin \mathcal{S}$. Then $H_{b,s} = \mathbb{P}[T_s < T_{\mathcal{S}\setminus s} | X(0) = b]$. Therefore on setting $i = s$ and $A = \mathcal{S} \setminus s$ we see that $H_{b,i} = \mathbb{P}[T_i^b < T_A^b]$ and $d(i, A) = \sum_{b \in \delta A} \Psi_B(b) H_{b,i}$. Recall that $\Psi_B(b)$ is the distribution in (5.8).

The elements of the matrix H are also weights used to represent the consensus value as a convex combination of the initial values of the stubborn agents (see [**14**],[**31**],[**18**]). To obtain a probabilistic interpretation of $d(i, A)$, say that $b \notin \mathcal{S}$ is attracted to $i \in \mathcal{S}$ if i is the first node in $\mathcal{S}$ encountered in a random walk starting at b. Then $d(i, A)$ measures the attractiveness of i to points adjacent to A.

REMARK. $(I - Q)$ is the grounded Laplacian L_g ([**40**],[**30**]) of the Markov chain graph.

5.3. A surrogate for F**.** The lesson of section 4 is that in contrast to the classical method, the greedy extension of a collection of small cardinality sets of high quality produces improved approximations of optimal sets of specified cardinality K. However there is no provision for finding better approximations if all of the sets have the same cardinality K. We must rely on trial and error swapping of single elements followed by evaluation of F. We developed a set valued function S based on a simplification of the upper bound in Corollary 1, that approximately (but satisfactorily) identifies highly optimal sets from among a collection of sets of the same size. Thus the use of S as an alternative to direct calculation of F is the basis of a promising heuristic for problem (3.2). The question we address in this section is how well can highly optimal sets be identified knowing only the number of uncovered edges, the degree of each node in $V \setminus A$ and the number of cut edges $|\partial A|$? The $d(i, A)$ terms are ignored, thus the influence of inaccessible nodes is accounted for only by the fact that the connected components have a large number of uncovered edges. Indeed obtaining the $\{d(i, A)\}_{i \notin A}$ would involve a computation comparable to calculating $F(A)$. Thus ranking sets by S is only a rough approximation of a ranking by F. Given the set A and assuming the uniform random walk the surrogate $S(A)$ is,

$$S(A) = 1 + \left(\frac{\sum_{b \in A^c} deg(b)}{|\partial A|} \right) \sum_{e \in \mathcal{E}} p_e \tag{5.17}$$

The quality of sets of the same cardinality can be expressed in terms of a normalized F value $\rho_F(A) = \frac{F_{max} - F(A)}{F_{max} - F_{min}}$ where $F_{max} = \max_{|A| \leq K} F(A)$ and $F_{min} = \min_{|A| \leq K} F(A)$. An analogous normalized S can be defined, $\rho_S(A) = \frac{S_{max} - S(A)}{S_{max} - S_{min}}$. The normalized values can be used to illustrate the correlation between highly optimal sets with respect to the objective function F and highly optimal sets with respect to the surrogate S. Figure 5 shows a plot of ρ_F versus ρ_S for the graph shown in Figure 2 for a collection of $33,649$ 5-element sets.

6. Conclusion

We considered the problem of maximizing the rate of convergence for the spread of consensus in the presence of stubborn agents in a network of fixed topology. We posed the optimization problem :determine the set of nodes A of fixed maximum cardinality that leads to the fastest rate of convergence to consensus. Here, following the random walk approach of [**6**], [**10**] the desired set minimizes $F(A)$, the sum of the first hitting times of random walks starting at a node outside the set A. In this paper we demonstrate that this choice of objective function has advantages that firstly enable the development of a potentially useful method of finding or approximating an optimal set and secondly allows us to gain some understanding of its structure. Using the facts that F is supermodular and that a vertex cover of the graph $\mathcal{G}$ is always a solution of the optimization problem for its cardinality, we developed a generalization of the classic greedy algorithm by constructing a class of optimal and near optimal small cardinality starter sets whose degree of optimality is defined in terms of a vertex cover. Previously a submodular function derived from F was used to prove that the performance ratio of this algorithm is at least $(1-1/e)$ (Proposition 1). The method is polynomial in N,the number of nodes and works well for moderate sized networks, but for very large networks the required computations of F become onerous. In section 5.3, we discussed a potential remedy for this. We presented heuristic evidence that an easy to calculate surrogate set function derived from an upper bound on F, can be used to identify highly optimal sets without direct computation of F, thus making it computationally cheaper to identify optimal and near optimal sets.

The use of first hitting time as a measure of the rate of convergence to consensus provides some insight into the topological characteristics of optimal and near optimal sets in ways that are analogous to the earlier results of [**18**] and [**30**] but unique to this work. The setting is an reversible irreducible random walk on $\mathcal{G}$, with transition matrix determined by the consensus model. In this paper, a recent equality of Basu et al [**5**] was used to derive an upper bound for $F(A)$ in terms of the bottle neck ratio of A^c , the number of edges in A^c non-adjacent to A (i.e. the number of uncovered edges) and quantities that measure the accessibility of a node $i \notin A$ to a random walk started at a node adjacent to A. In fact we show (see section 5.2) that these accessibility measures can be written in terms of components of the eigenvector associated with the smallest eigenvalue of an augmented Laplacian. The work of [**14**] shows that the probabilities in equation (5.13) appear as weights in the representation of the consensus value.

Appendix A. Rate of convergence to consensus and first hitting time

To motivate the optimization problem whose solution will define an effective spreader, we will discuss the link between the (rate of convergence) to consensus in a network containing a set A of leader (stubborn) agents and the mean first hitting time (or arrival time) to A of a random walk on the underlying graph.

The information state of a network at a discrete time n; $n = 0, 1, \cdots$ is represented by a vector $U_n \in \mathbb{R}^N$ whose ith component is a real number u_n^i that represents the information state of node i at time n. The dynamics of the consensus process in the network for $n \geq 1$ is given by:

$$u_n^i = \begin{cases} c \geq 0, & \text{if } i \in A \\ \sum_{j \sim i, j \notin A} P(i,j) u_{n-1}^j + c \sum_{j \sim i, j \in A} P(i,j) & i \notin \text{A} \end{cases} \tag{A.1}$$

The initial value U_0 at time $n = 0$ is,

$$u_0^i = \begin{cases} c & \text{if } i \in A \\ 0 & \text{if } i \notin \text{A} \end{cases} \tag{A.2}$$

Although these values are assigned for convenience there is no loss of generality in doing so because the equations are linear and the vector $c\mathbf{1}$ is a solution of equation (A.1).

Equation (A.1) states that the information value at node i changes to the information value at node j if a random walker steps from i to j in a single time step. In particular, if $j \in A$, the value at i changes to c and is unchanged thereafter. Thus u_n^i is the expected information value at i at time n after a single random walk step. At this point it will be convenient to introduce a variable that only tracks the dynamics in A^c. Thus we let

$$u_n^i = \begin{cases} 0 & \text{if } i \in A \\ x_n^i & \text{if } i \notin \text{A} \end{cases} \tag{A.3}$$

If x_n is the vector whose ith component is x_n^i, then the consensus equations (A.1), (A.2) become:

$$x_{n+1} = P_A x_n + c R_A \mathbf{1}, \;\; x_0 = 0 \tag{A.4}$$

The vector $\mathbf{1}$ is a column vector of $N - |A|$ ones. Here $P_A(i,j) = P(i,j)$ when $i,j \in A^c$, and $R_A(i,j) = P(i,j)$, when $i \in A^c, j \in A$. Since the transition matrix P (see section 2) is stochastic, $R_A = I - P_A$. Here I is the $N - |A| \times N - |A|$ identity matrix. From (A.4) it is then clear that $x_1 = cR_A\mathbf{1} = c(I - P_A)\mathbf{1}$ and it is not hard to establish by mathematical induction that

$$x_n = c(I - P_A^n)\mathbf{1}, \tag{A.5}$$

so that the substochastic property of P_A implies that $x_n \to x^*$ as $n \to \infty$, with $x^* = c\mathbf{1}$. In this sense, the information value common to nodes at A spreads to the rest of the network.

We can define the rate of convergence to consensus to be the rate of convergence of the ℓ_1 difference $||x^* - x_n||_1$ to 0. The following equation shows that the rate is controlled by the tail of the distribution of the first hitting time to A.

$$||x^* - x_n||_1 = \mathbb{P}[T_A > n] \tag{A.6}$$

The proof of (A.6) will be presented shortly. It uses standard results in the theory of absorbing Markov chains. For finite irreducible Markov chains, it is known that the right hand side decays exponentially in n. For an arbitrary initial distribution μ on the states of the chain, ([**34**] Chapter 3 and [**3**] Chapter 2),

$$\mathbb{P}_\mu[T_A > n] \leq \exp\left(-\left\lfloor \frac{n}{\lceil et_A^* \rceil} \right\rfloor\right) \tag{A.7}$$

where $t_A^* = \max_{i \in V} E_i[T_A]$. Now note that,

$$\max_{i \in V} \mathbb{E}_i[T_A] \leq F(A) = \sum_{i \in V} \mathbb{E}_i[T_A] \leq (N - |A|) \max_{i \in V} \mathbb{E}_i[T_A]$$

Thus equation (A.7) shows that the rate of convergence is determined (up to a constant) by F. A reasonable strategy then for finding the set that maximizes the rate of convergence to consensus and hence finding an effective spreader is to seek a solution to the following optimization problem:

$$\min_{A \subset V, |A| = K} F(A) \tag{A.8}$$

LEMMA A.1. *Let $\mathbb{P}$ be the probability distribution associated with an irreducible Markov chain with transition matrix P and suppose $A \subset V$ is a non-empty subset. If x_n is the vector at time n of the consensus process* (A.4)*, with limiting vector x^*, then*

$$||x^* - x_n||_1 = \mathbb{P}[T_A > n] \tag{A.9}$$

PROOF. From equation (A.5) we know that $||x^* - x_n||_1 = c||P_A^n\mathbf{1}||_1$. Now the ith component of $P_A^n\mathbf{1}$ is $\sum_{j \in A^c} P^n(i,j) = P_i[T_A > n]$, the probability that starting at i, the random walker has not arrived at A after n steps. Thus to compute the ℓ_1 norm we must have,

$$||x^* - x_n||_1 = \sum_{i \in A^c} P_i[T_A > n] = \mathbb{P}[T_A > n]$$

□

Acknowledgments

The author acknowledges the late Chris Dabrowski for providing the data and thanks Elie Al Hajjar for help in the preparation of Figure 3.

References

[1] Daron Acemoğlu, Giacomo Como, Fabio Fagnani, and Asuman Ozdaglar, *Opinion fluctuations and disagreement in social networks*, Math. Oper. Res. **38** (2013), no. 1, 1–27, DOI 10.1287/moor.1120.0570. MR3029476

[2] Rafig Agaev and Pavel Chebotarev, *On the spectra of nonsymmetric Laplacian matrices*, Linear Algebra Appl. **399** (2005), 157–168, DOI 10.1016/j.laa.2004.09.003. MR2152415

[3] D.Aldous, J.Fill, *Reversible Markov Chains and Random Walks on Graphs*, available at `https://www.stat.berkeley.edu/~aldous/RWG/book.html`

[4] Ashwin Arulselvan, Clayton W. Commander, Lily Elefteriadou, and Panos M. Pardalos, *Detecting critical nodes in sparse graphs*, Comput. Oper. Res. **36** (2009), no. 7, 2193–2200, DOI 10.1016/j.cor.2008.08.016. MR2586486

[5] Riddhipratim Basu, Jonathan Hermon, and Yuval Peres, *Characterization of cutoff for reversible Markov chains*, Ann. Probab. **45** (2017), no. 3, 1448–1487, DOI 10.1214/16-AOP1090. MR3650406

[6] V.S. Borkar, J. Nair, N. Sanketh, *Manufacturing Consent*, 48th Annual Allerton Conference, Allerton House, UIUC, Illinois, September 2010, pp. 1550-1555

[7] B. Caskurlu, K.Subramani, *On Partial Vertex Cover on Bipartite Graphs and Trees*, 2013 `arXiv:1304.5934v1`

[8] Stephen Boyd, Arpita Ghosh, Balaji Prabhakar, and Devavrat Shah, *Randomized gossip algorithms*, IEEE Trans. Inform. Theory **52** (2006), no. 6, 2508–2530, DOI 10.1109/TIT.2006.874516. MR2238556

[9] Pavel Chebotarev and Rafig Agaev, *Forest matrices around the Laplacian matrix*, Linear Algebra Appl. **356** (2002), 253–274. Special issue on algebraic graph theory (Edinburgh, 2001). MR1945250

[10] Andrew Clark, Basel Alomair, Linda Bushnell, and Radha Poovendran, *Minimizing convergence error in multi-agent systems via leader selection: a supermodular optimization approach*, IEEE Trans. Automat. Control **59** (2014), no. 6, 1480–1494, DOI 10.1109/TAC.2014.2303236. MR3225224

[11] A. Clark, L. Bushnell, R. Poovendran, *Leader Selection for Minimizing Convergence Error in Leader-Follower Systems:A Supermodular Optimization Approach*, 10th International Symposium Modeling and Optimization in Mobile, Ad-Hoc and Wireless Networks (WiOpt), May 2012, pp. 111-115

[12] Andrew Clark, Basel Alomair, Linda Bushnell, and Radha Poovendran, *On the structure and computation of random walk times in finite graphs*, IEEE Trans. Automat. Control **64** (2019), no. 11, 4470–4483. MR4030773

[13] Thomas H. Cormen, Charles E. Leiserson, Ronald L. Rivest, and Clifford Stein, *Introduction to algorithms*, 3rd ed., MIT Press, Cambridge, MA, 2009. MR2572804

[14] Giacomo Como and Fabio Fagnani, *From local averaging to emergent global behaviors: the fundamental role of network interconnections*, Systems Control Lett. **95** (2016), 70–76, DOI 10.1016/j.sysconle.2016.02.003. MR3514363

[15] C.Dabrowksi, K. Mills, *Using realistic factors to simulate catastrophic congestion events in a network*, Computer Communications, Vol. 112, November 2017, pp. 93-108

[16] M.H.DeGroot, *Reaching A consensus*, J. Am. Stat. Assoc., Vol. 69, No. 345, pp.118-121, 1974

[17] Duanbing Chen, Linyuan Lu, Mingshen Shang,Yi-Cheng Zhang,Tao Zhou, *Identifying influential nodes in complex networks*, Physica A,Vol. 391, No. 4, 2012

[18] Javad Ghaderi and R. Srikant, *Opinion dynamics in social networks with stubborn agents: equilibrium and convergence rate*, Automatica J. IFAC **50** (2014), no. 12, 3209–3215, DOI 10.1016/j.automatica.2014.10.034. MR3284157

[19] Christopher Griffin, Sarah Rajtmajer, Anna Squicciarini, and Andrew Belmonte, *Consensus and information cascades in game-theoretic imitation dynamics with static and dynamic network topologies*, SIAM J. Appl. Dyn. Syst. **18** (2019), no. 2, 597–628, DOI 10.1137/16M109675X. MR3934101

[20] F. Hunt, *The Structure of Optimal and Near Optimal Sets in Consensus Models*, NIST Special Publication 500-303, August 2014, `arXiv:1408.4364` (online version)

[21] F.Hunt, *An Algorithm for Identifying Optimal Spreaders in a Random Walk Model of Network Communication*, Journal of Research of the National Institute of Standards and Technology, Volume 121 (2016) http://dx.doi.org/10.6028/jres.121.008

[22] Victor P. Il′ev, *An approximation guarantee of the greedy descent algorithm for minimizing a supermodular set function [MR1680299 (2000a:90053)]*, Discrete Appl. Math. **114** (2001), no. 1-3, 131–146, DOI 10.1016/S0166-218X(00)00366-8. Discrete analysis and operations research. MR1865587

[23] Dieter Jungnickel, *Graphs, networks and algorithms*, 4th ed., Algorithms and Computation in Mathematics, vol. 5, Springer, Heidelberg, 2013. MR2986014

[24] John G. Kemeny and J. Laurie Snell, *Finite Markov chains*, Springer-Verlag, New York-Heidelberg, 1976. Reprinting of the 1960 original; Undergraduate Texts in Mathematics. MR0410929

[25] David A. Levin, Yuval Peres, and Elizabeth L. Wilmer, *Markov chains and mixing times*, American Mathematical Society, Providence, RI, 2009. With a chapter by James G. Propp and David B. Wilson. MR2466937

[26] Van Sy Mai and Eyad H. Abed, *Optimizing leader influence in networks through selection of direct followers*, IEEE Trans. Automat. Control **64** (2019), no. 3, 1280–1287. MR3922094

[27] G. L. Nemhauser and L. A. Wolsey, *Best algorithms for approximating the maximum of a submodular set function*, Math. Oper. Res. **3** (1978), no. 3, 177–188, DOI 10.1287/moor.3.3.177. MR506656

[28] G. L. Nemhauser, L. A. Wolsey, and M. L. Fisher, *An analysis of approximations for maximizing submodular set functions. I*, Math. Programming **14** (1978), no. 3, 265–294, DOI 10.1007/BF01588971. MR503866

[29] R. Olfati-Saber, J. Fax, R.M. Murray, *Consensus and cooperation in networked multi agent systems*, Proceedings IEEE, Vol. 95, January 2007, pp. 215-233

[30] M. Pirani, S. Sundaram, *Spectral Properties of the Grounded Laplacian Matrix with Applications to Consensus in the Presence of Stubborn Agents*, Proceedings of ACC2014, the 33rd, American Control Conference, Portland OR, 2014

[31] M.Pirani, *The Spectral Properties of the Grounded Laplacian Matrix*, thesis University of Waterloo, Department of Electrical and Computer Engineering, Waterloo, Ontario, 2014

[32] Mohammad Pirani and Shreyas Sundaram, *On the smallest eigenvalue of grounded Laplacian matrices*, IEEE Trans. Automat. Control **61** (2016), no. 2, 509–514. MR3454759

[33] J. W. Pitman, *Occupation measures for Markov chains*, Advances in Appl. Probability **9** (1977), no. 1, 69–86, DOI 10.2307/1425817. MR433600

[34] S. Roch, `http://www.math.wisc.edu/~roch/mdp/index.html` (accessed Oct. 30, 2020)

[35] Guodong Shi, Kin Cheong Sou, Henrik Sandberg, and Karl Henrik Johansson, *A graph-theoretic approach on optimizing informed-node selection in multi-agent tracking control*, Phys. D **267** (2014), 104–111, DOI 10.1016/j.physd.2013.07.014. MR3132393

[36] Daniel A. Spielman and Shang-Hua Teng, *Nearly linear time algorithms for preconditioning and solving symmetric, diagonally dominant linear systems*, SIAM J. Matrix Anal. Appl. **35** (2014), no. 3, 835–885, DOI 10.1137/090771430. MR3228466

[37] J. Tsitsiklis, PhD Thesis, Massachusetts Institute of Technology, 1984

[38] Alex Olshevsky and John N. Tsitsiklis, *Convergence speed in distributed consensus and averaging [reprint of MR2480125]*, SIAM Rev. **53** (2011), no. 4, 747–772, DOI 10.1137/110837462. MR2861265

[39] Zengfu Wang, Bill Moran, Xuezhi Wang, and Quan Pan, *Approximation for maximizing monotone non-decreasing set functions with a greedy method*, J. Comb. Optim. **31** (2016), no. 1, 29–43, DOI 10.1007/s10878-014-9707-3. MR3440244

[40] Weiguo Xia and Ming Cao, *Analysis and applications of spectral properties of grounded Laplacian matrices for directed networks*, Automatica J. IFAC **80** (2017), 10–16, DOI 10.1016/j.automatica.2017.01.009. MR3638051

[41] E.Yildiz, A.Ozdaglar, D.Acemoglu, A.Saberi, A.Scaglione, *Binary Opinion Dynamics with Stubborn Agents* ACM Transactions on Economics and Computations, Vol. 1, No. 4, Article 19, December 2013

Applied and Computational Mathematics Division, National Institute of Standards and Technology, Gaithersburg, Maryland 20899

Email address: `fern.hunt@nist.gov`

Contemporary Mathematics
Volume **759**, 2020
https://doi.org/10.1090/conm/759/15268

Peer-Led Team Learning and its effect on mathematics self-efficacy and anxiety in a developmental mathematics course

Nadia Monrose Mills, Angelicque Tucker Blackmon, Camille McKayle, Robert Stolz, and Sandra Romano

Abstract. Peer-Led Team Learning (PLTL) was implemented in the developmental mathematics courses at an open enrollment historically black college and university (HBCU). The implementation of PLTL, as a required component in these courses, resulted in pass rates increasing from below 50% to almost 80%. With this success, this study seeks to explore the impact of PLTL on mathematics self-efficacy and mathematics anxiety, two non-cognitive factors correlated to student achievement. Preliminary findings from the Mathematics Self-Efficacy and Anxiety Questionnaire (MSEAQ) [**11**] for the spring 2017 semester are shared. Inferential statistical analysis shows a slight increase in mathematics self-efficacy and a slight decrease in mathematics anxiety. Although these differences are not statistically significant, they are promising. This analysis resulted in increased data collection efforts and programmatic improvements. These improvements included changes in training materials for peer leaders to address mathematics efficacy and anxiety along with revisions to the PLTL curriculum.

1. Introduction

There is a national shortage in the number of underrepresented groups attaining STEM degrees and entering the workforce. However, diversity in the workforce is necessary to keep American companies competitive in a global economy [**12**]. Historically Black Colleges and Universities (HBCUs) have been successful in graduating underrepresented groups with STEM degrees, at the undergraduate and graduate levels [**14**], [**13**].

At the University of the Virgin Islands (UVI), an HBCU in the Caribbean, through targeted interventions, the number of STEM majors has increased along with STEM student retention and persistence rates. STEM programs and activities implemented not only seek to improve students' content knowledge but also improve other non-cognitive factors that have been proven to affect student success in STEM. The purpose of this paper is to describe Peer-Led Team Learning (PLTL) in our developmental mathematics courses and preliminary results of its impact on mathematics self-efficacy and mathematics anxiety.

2020 *Mathematics Subject Classification.* Primary 97D40, 97D60.

2. Background and context

Founded in 1962, the University of the Virgin Islands (UVI) is a public, open admissions, HBCU and a Land and Sea Grant institution with an enrollment of approximately 2200 students across two campuses: one on St. Thomas and one on St. Croix. This enrollment was severely impacted by the passage of two category five hurricanes in Fall, 2017, and enrollment fell to approximately 2,000 FTE students.

UVI primarily serves students from a unique cultural, ethnic, and socioeconomic background in the United States Virgin Islands. The US Virgin Islands is a territory of the United States, 1100 miles southeast of Miami, consisting of three major islands: St. Thomas, St. Croix, and St. John. The population of about 105,000 is 76% Black, 17% Hispanic, and 15% White. Culturally, the Virgin Islands are diverse, including immigrants from throughout the English-speaking Caribbean, Puerto Rico, Haiti, and the Dominican Republic. Approximately 69% of UVI students are African-American and 7% are Hispanic; 94% are US citizens or permanent residents; most students are female (70%) and most are commuters. Also, 56% of undergraduates receive Pell grants.

As the only institution of higher learning in the Virgin Islands, UVI is constantly working towards offering well-designed programs, activities and courses. The most critical courses for students beginning a career in science, technology, engineering or mathematics are the courses in the first-year college-level mathematics sequence. These courses can act as gatekeeper courses (Fullilove and Treisman 1990) and can adversely affect students' progress. About 60% of our freshmen STEM students do not score high enough on our mathematics placement exams to enroll in first-year College Algebra courses. As a result, these students are unable to register for the first-year general chemistry and computer science courses.

Beginning in 1999, UVI secured two funded projects from the National Science Foundation (NSF) to create a holistic approach to increasing the retention, persistence, and graduation rates of STEM students, involving curricular, co-curricular and student life interventions. For these projects, retention was defined as the percentage of full-time freshmen degree seeking students who returned to UVI for the second year. Persistence was defined as the percentage of full-time students that have remained enrolled, even if they change institutions. UVI used the Integrated Post-Secondary Education Data System (IPEDS) graduation rate which is the percentage of full-time students who graduate within four to six years to define persistence. These funded programs led to the creation and integration of Peer-Led Team Learning, a research based active learning course supplement into the Introduction to Algebra Concepts I & II courses which comprises the developmental mathematics course sequence.

After the implementation of these funded projects, the number of STEM majors has increased 45% from 2001 to 2019. There was also a 62% increase in the numbers of STEM graduates, from 26 Bachelor's degrees in 2001 to 42 in 2019. Retention rates for STEM majors have improved, with the 2015-2016 cohort at 91%, versus UVI's 75% overall rate. The persistence rate has also increased for STEM majors. It is at approximately 65%, versus UVI's overall 58% for 2014-2016. Graduation rates for STEM majors currently hover near 30%, versus UVI's overall six-year graduation rates of 22% to 26% (2010 cohort). Of the 35 STEM majors graduating from UVI in 2015, approximately 20% went on to PhD programs in STEM, compared to 14% of STEM Bachelor's degree recipients nationally who entered STEM PhD programs.

However, continued intentional analysis of data from UVI's Office of Institutional Research and Planning shows that there remains an equity gap in terms of retention and graduation rates for those students who struggle early in their college careers. Data show that students with GPAs below 2.3 are retained at a much lower rate than those with a higher GPA. Students with GPAs below 2.3 also have a much lower average 6-year graduation rate. Their graduation rate is just 6.16%, compared to 41% for those students with a higher GPA. This suggests that early success is a predictor of persistence and academic attainment, hence the need for course intervention programs like PLTL.

PLTL favorably impacts students who struggle early in their college careers is Peer-Led Team Learning (PLTL). In this paper, preliminary findings on PLTL's impact on mathematics self-efficacy and mathematics anxiety for students in the non-credit bearing developmental mathematics courses are shared.

2.1. Research context. In this section, a brief description of the research on mathematics self-efficacy, mathematics anxiety, and Peer-Led Team Learning will be given. Mathematics self-efficacy has been described as the judgement and assessment of one's own ability to successfully learn mathematics and execute a mathematical task [**17**]. This construct has been associated with college students' mathematics achievement. In a study of college freshmen, Hall and Ponton (2002),[**6**], explored the differences between students enrolled in a developmental mathematics course and those enrolled in a calculus course. Students in the developmental course had lower mathematics self-efficacy than students in the Calculus course. In another study with university students in a statistics course, Azar and Mahmoudi (2014) [**2**], found that students with higher mathematics self-efficacy earned higher test scores. These studies among many others are aligned with Bandura's self-efficacy theory [**3**], that achievement in mathematical skill is the greatest source of mathematics self-efficacy. To this end, freshmen students in developmental mathematics courses will most likely have decreased mathematics self-efficacy due to their past failures in mathematics [**19**]. Therefore, students who previously performed well in mathematics have higher mathematics self-efficacy scores.

Mathematical anxiety is also a factor that determines success in mathematics. Mathematics anxiety is described as the tension and anxiety one experiences from interactions with numerical manipulations in formal and informal settings [**16**]. Azar and Mahmoudi (2014),[**2**] in a study on statistics students at the university level, found that as mathematics anxiety increased, students' scores decreased. The low performance in the classroom and on tests for students who experienced mathematics anxiety can be attributed to their difficulty accessing information [**7**] likely due to the impact that anxiety has on working memory [**1**].

Peer-Led Team Learning (PLTL) is a research based active learning course component that has been shown to increase students' performance and success rate in science and mathematics courses [**5**], [**4**]. This supplemental course component, called PLTL Workshop sessions, meets weekly in small groups (6-8 students) to work on challenging problems facilitated by an advanced student, called a peer leader [**5**]. Peer leaders are ideally students who have completed the course with a grade B or higher. PLTL began with Chemistry courses among a few universities but has expanded to other schools and STEM subjects (e.g. Mathematics, Biology,

Physics, Computer Science) after the success of these workshops on the course pass rate and positive impacts on students and peer leaders became popular [**10**], [**18**], [**8**], [**15**], [**9**]. There are seven critical components that each PLTL program must include to maintain implementation fidelity [**5**], [**4**]. One example component is that workshops must be required, occurs regularly, and coordinated with the course. A more thorough description of the seven critical components will be discussed in the Methods section.

2.2. Hypothesis. There were two main hypotheses in the current study. First, we expected that the PLTL program would lead to increases in students' mathematics self-efficacy. Second, we predicted that PLTL would decrease students' mathematics anxiety.

Method

This section describes the sample population and methodology used to study the impact of PLTL on students' mathematical self-efficacy and mathematics anxiety.

2.3. Participants. UVI students attending developmental mathematics PLTL Workshop sessions voluntarily participated in this research study. Students were STEM and non-STEM majors that scored below a 520 on the SAT or did not meet the minimum requirement to take the first college-level mathematics course based on their performance on a placement test. About 60% of our STEM majors are placed into the developmental mathematics program. The total number of students in this mathematics program ranged from as low as 223 students to about 396 students over the past several years. During Spring 2017, when data for this study were collected, 141 students responded to the pre-survey and 71 students responded to the post survey. The smaller number of participants in the post-survey was not due to lower enrollment at the end of the semester but rather a decrease in post-course survey responses.

This developmental mathematics program consists of two courses, Introduction to Algebra Concepts I & II. Part I covers pre-algebra topics such as integers, order of operations, solving linear equations, rational numbers and polynomials. Part II covers primarily linear functions, systems of linear equations, and quadratic functions. Both parts are four-credit hours but since they are non-credit bearing, students earn a P for passing or NP for not passing. These courses are taught primarily by adjunct faculty (about 80% coverage) and are coordinated by two full time faculty members, one on each campus. These courses are standardized. That is, all sections follow the same syllabus, grading scheme, and administer coordinated midterm exams and final exams. The coordinators provide professional development to the faculty and creates all exams. Students must earn a 700 out of 1000 total points to pass.

PLTL was first introduced in the developmental mathematics courses in 2012 as a pilot to improve the pass rates which were well below 50%. The sections in Part I that included PLTL during the pilot program saw a 70% pass rate. Due to the success of the pilot program, PLTL became mandatory for all students in the foundations courses and was institutionalized. Workshops were scheduled to serve over 300 students across both campuses. On the St. Thomas campus, there can be at most 22 Workshop sessions shared among 12-14 peer leaders.

The structure of UVI's PLTL program ensures that the model follows the seven critical components fidelity ([**4**],[**5**]). First, workshop sessions meet weekly for 50 minutes and are a required component of the course. Second, a mathematics faculty member, one on each campus, provides oversight to the program and designs the workshop activities, and selects and train the leaders. Third, peer leaders attend a paid, mandatory meeting once a week for training on the mathematics content and on facilitation techniques. Fourth, the workshop activities are aligned with the curriculum. Fifth, workshop sessions are comprised of 6-8 students assigned to one peer leader who meets students at the same location each week. Sixth, the extra hour was officially scheduled, with peer leaders paid through university funds. The faculty coordinator also received release time to train leaders and for continued enrichment of the curriculum. Seventh, through facilitation techniques students learned to work effectively as a team. As a result of implementing a PLTL program that fulfills these critical components, there have been increases in the course's pass rates in subsequent years from below 50% to almost 80%.

2.4. Procedure. Pre- and post-surveys were distributed digitally through SurveyMonkey. The pre-survey was administered during the first two weeks of the semester and the post-survey was administered during the final two weeks of the semester. The survey link was made available to students through their Blackboard course page. The data manager and educational researcher drafted a message that was then sent to all students in the developmental mathematics courses. The instrument that was used to measure mathematics efficacy and anxiety is The Mathematics Self-Efficacy and Anxiety Questionnaire (MSEAQ; May, 2009). MSEAQ is a 29-item survey with two subscales. The Mathematics Self-Efficacy subscale has 14 questions and a Cronbach's coefficient α of 0.93 (May 2009). The Mathematics Anxiety subscale has 15 questions also with a Cronbach's coefficient α of 0.93. The overall Cronbach's coefficient α of the MSEAQ is 0.96 (May, 2009). A Likert style scoring system was used. Students responded to a Likert scaled survey with responses ranging from 1 to 5 (1=never; 2=seldom; 3=sometimes; 4=often; and 5=usually).

Results

To determine the impact of PLTL on students' mathematics efficacy and anxiety, descriptive and inferential statistical analyses were completed on pre- and post-survey responses. Descriptive statistical analysis, using SPSS, was performed to determine mathematics self- efficacy and mathematics anxiety means and standard deviations for students participating in 13-weeks of PLTL Workshop sessions. Among the descriptive variables for entering freshman students were gender, ethnicity, average high school mathematics grade, high school mathematics courses taken, college GPA, and college classification. These descriptive variables were not used in this analysis. Researchers performed t-tests on the mathematics efficacy and anxiety scores between the pre-surveys and post-surveys. Change was considered significant if the p-value was less than 0.05. Inferential statistical analysis was done to determine whether to accept the null hypothesis. The null hypothesis, H_0, stated that there will be no difference between the pre-test scores and post-test scores. The alternative hypothesis stated that there will be a statistically significant difference between the pre-survey and post-survey scores indicating statistically significant increases in students' mathematics self-efficacy and decrease in mathematics anxiety.

2.5. Mathematics self-efficacy. Table 1 shows the pre-post means for mathematics efficacy using the MSEAQ (May 2009). Table 1 shows pre-post increases for all survey questions except question 10. This response indicates that most students sometimes or often believe they will be able to use mathematics in their future careers when needed.

TABLE 1. Pre-Post Comparison of Mathematics Self-Efficacy

	Pre-Mean	Post Mean
1.I feel confident enough to ask questions in my mathematics class	4.00	4.13
4.I believe I can do well on a mathematics test.	3.65	3.69
9.I believe I am the kind of person who is good at mathematics	2.94	3.01
10.I believe I will be able to use mathematics in my future career when needed.	3.71	3.70
12.I believe I can understand the content in a mathematics course.	3.65	3.71
13.I believe I can get an A when I am in a mathematics course	3.16	3.37
16.I believe I can learn well in a mathematics course.	3.75	3.76
19.I feel confident when taking a mathematics test.	2.97	3.27
20.I believe I am the type of person who can do mathematics.	3.37	3.38
21.I feel that I will be able to do well in future mathematics courses.	3.35	3.54
23.I believe I can do the mathematics in a mathematics course	3.55	3.59
28.I believe I can think like a mathematician.	2.53	2.56
29.I feel confident when using mathematics outside of school	3.12	3.14

Table 2 shows the pre-post means and standard deviations on the mathematics self-efficacy subscale. The number of respondents for the pre-survey was 141. There were 71 respondents for the post-survey. Because of the low post-survey response rate and difficulty matching participants from pre- and post-surveys, an independent-samples t-test was conducted to determine whether mathematics self-efficacy was higher after the PLTL program. Results from Table 2 shows no statistically significant differences between the pre-test and post-test scores.

The PLTL component within the course made no statistical difference in students' mathematics self-efficacy. However, the pre-post means show a slight increase from 3.36 (pre-survey) to 3.46 (post-survey). These means reveal that students responded between sometimes (3) and often (4) for each item on the self-efficacy scale. These means also reveal that students perceive they sometimes or often perceive

TABLE 2. Independent T-Test of Mathematics Self-Efficacy

Testing Conditions						95% CI for Mean Difference		
Pre Test			Post Test					
M	SD	n	M	SD	n		t	df
3.36	.74	141	3.46	.76	.71	−.31, 12	−.867	210

that they have confidence in their mathematics abilities or believe that they will do well in mathematics courses or on tests or exams. While no statistical difference was found between the two groups it should be noted that the post-survey mean score increased

2.6. Mathematics anxiety. Table 3 shows the pre-post means and standard deviations for mathematics anxiety subscale using the MSEAQ (May 2009). The number of respondents for the pre-survey was 141. There were 71 respondents for the post-survey. Table 3 shows pre-post decreases on only 5 out of the 15 mathematics anxiety questions. The questions that show an increase in mathematics anxiety are questions 3, 11, 15, 18, and 26. Question 11 which states, "I feel stressed when listening to mathematics instructors in class," shows the greatest increase. Question 26, "I get nervous while taking a mathematics test" measures mathematics test anxiety and also shows larger pre-post increase than other question responses. Feeling stressed while listening to their instructor and mathematics testing anxiety are contributing to the overall mean of the anxiety subscale than most other factors.

Table 3 shows the pre-post means and standard deviations on the mathematics anxiety subscale. The number of respondents for the pre-survey is 141, and there were 71 respondents for the post-survey. Because of the low post-survey response rate and difficulty matching participants from pre- and post-surveys, an independent-samples t-test was conducted to test the hypothesis that mathematics anxiety would decrease after exposure to the PLTL program. Results from Table 3 shows no statistically significant difference in the pre/post means for mathematics anxiety.

The pre- and post-means show a slight increase from 2.57 (pre-survey) to 2.60 (post-survey). This increase reveals that students responded between seldom (2) and sometimes (3) for each item on the anxiety scale. These means indicate that students are seldom or sometimes worried or nervous about their performance on mathematics tests or class assignments. However, questions 2, 8, and 26 show higher pre-test responses indicating that students were more nervous or worried when taking mathematics tests. While no statistical difference was found between the two groups it should be noted that the post-test mean score did not decrease.

3. Discussion

There is empirical evidence that Peer-Led Team Learning in the developmental mathematics program at the University of the Virgin Islands is moving towards accomplishing the goal of increasing students' mathematics efficacy and decreasing mathematics anxiety. Preliminary analysis of only the spring 2017 data did not show statistically significant results but suggests promising trends. There was a

TABLE 3. Mathematics Anxiety Pre/post comparisons

	Pre-Mean	Post Mean
2. I get tense when I prepare for a mathematics test.	3.18	2.93
3. I get nervous when I have to use mathematics outside of school.	2.14	2.25
5. I worry that I will not be able to use mathematics in my future career when needed.	2.12	2.17
6. I worry that I will not be able to get a good grade in my mathematics course.	2.70	2.76
8. I worry that I will not be able to do well on mathematics tests.	2.92	2.64
11. I feel stressed when listening to mathematics instructors in class.	2.20	2.55
14. I get nervous when asking questions in class.	2.00	2.06
15.Working on mathematics homework is stressful for me.	2.66	2.78
17. I worry that I do not know enough mathematics to do well in future mathematics courses.	2.64	2.61
18. I worry that I will not be able to complete every assignment in a mathematics course.	2.50	2.66
22. I worry I will not be able to understand the mathematics.	2.63	2.51
24. I worry that I will not be able to get an A in my mathematics course.	2.82	2.76
25. I worry that I will not be able to learn well in my mathematics course	2.47	2.55
26. I get nervous when taking a mathematics test	2.89	3.01
27. I am afraid to give an incorrect answer during my mathematics class	2.76	2.78

TABLE 4. Independent t-test for Mathematics Anxiety

Testing Conditions						95% CI for Mean Difference		
Pre Test			Post Test					
M	SD	n	M	SD	n		t	df
2.57	.78	131	3.60	.89	67	$-.27, .21$	$-.243$	196

slight increase in students' mathematics self-efficacy after 13 weeks of PLTL. Although there was an increase in students' overall mathematics anxiety, there were decreases on some of the items. Findings from the mathematics anxiety subscale shows higher pre-survey scores for items pertaining to taking mathematics test. This indicates that students in the developmental mathematics courses are most anxious about testing.

What are the implications of these results? Firstly, data collection efforts were improved to ensure that the sample sizes from the pre-survey to the post-survey were more consistent. Initially, surveys were distributed through Blackboard. Because the initial response rate was low, researchers added time during PLTL Workshop sessions to complete the pre- and post-surveys. Secondly, the peer-leader training program was enhanced by including training modules on theories of intelligence (Dweck, 2006), and mathematics anxiety. Peer leaders completed readings and engaged in activities during their weekly training sessions. Training activities included growth mindset and fixed mindset topics. The PLTL curriculum was also enhanced to include writing activities. Future PLTL sessions will incorporate verbal feedback designed to decrease students' mathematics anxiety and increase their mathematics self-efficacy and growth mindset.

4. Future directions

Peer-Led Team Learning has improved the passing rates in the developmental courses at the University of the Virgin Islands. Before PLTL, pass rates were below 50% and increased to almost 80% in some semesters. However, to determine the impact of PLTL on mathematics self-efficacy and mathematics anxiety, an increase in sample size is needed. This increase will occur as we compile data over six semesters. In addition, continuous improvement on our PLTL model is needed. For example, we need to remain vigilant implementing the seven critical components of the PLTL model. Finding creative ways to increase PLTL Workshop sessions from 50 minutes to 120 minutes is one strategy to improve our alignment to the model. Constant revision of the Workshop curriculum and the peer leader training materials will also improve our alignment to the model and keep students and peer-leaders interested and motivated [**4**]. As with any program, constant improvements will sustain the effectiveness of PLTL on students' content knowledge along with increasing their mathematics efficacy and decreasing their mathematics anxiety.

References

[1] Mark H Ashcraft and Jeremy A Krause, *Working memory, math performance, and math anxiety*, Psychonomic bulletin & review **14** (2007), no. 2, 243–248.

[2] Firouzeh Sepehrian Azar and Limou Mahmoudi, *Relationship between mathematics, self-efficacy and students' performance in statistics: the meditational role of attitude toward mathematics and mathematics anxiety*, Journal of Educational Sciences & Psychology **4** (2014), no. 1.

[3] Albert Bandura, *Self-efficacy. encyclopedia of human behavior (vol. 4, pp. 71-81)*, 1994.

[4] Leo Gafney and Pratibha Varma-Nelson, *Peer-led team learning: Evaluation, dissemination, and institutionalization of a college level initiative*, vol. 16, Springer Science & Business Media, 2008.

[5] Cracolice M. S. Kampmeier J. A. Roth V. Strozak V. S. Varma-Nelson P. Gosser, D. K., *Peer-led team learning: A guidebook*, Prentice Hall Upper Saddle River, NJ, 2001.

[6] J Michael Hall and Michael K Ponton, *Mathematics self-efficacy of college freshman.*, Journal of Developmental Education **28** (2005), no. 3, 26.

[7] Claudia L Hines, Nina W Brown, and Steve Myran, *The effects of expressive writing on general and mathematics anxiety for a sample of high school students*, Education **137** (2016), no. 1, 39–45.

[8] Susan Horwitz, Susan H Rodger, Maureen Biggers, David Binkley, C Kolin Frantz, Dawn Gundermann, Susanne Hambrusch, Steven Huss-Lederman, Ethan Munson, Barbara Ryder, et al., *Using peer-led team learning to increase participation and success of under-represented groups in introductory computer science*, ACM SIGCSE Bulletin **41** (2009), no. 1, 163–167.

[9] Yuanyuan Kang, Lisa Matsell, and Mitsue Nakamura, *Implementation of pltl in a freshman biology course at the university of houston-downtown*, Conference Proceedings of the Peer-Led Team Learning International Society, 2012.
[10] Kenneth S Lyle and William R Robinson, *A statistical evaluation: Peer-led team learning in an organic chemistry course*, 2003.
[11] Diana Kathleen May, *Mathematics self-efficacy and anxiety questionnaire*, Ph.D. thesis, University of Georgia, 2009.
[12] Engineering National Academies of Sciences, Medicine, et al., *Minority serving institutions: America's underutilized resource for strengthening the stem workforce*, National Academies Press, 2019.
[13] Emiel W Owens, Andrea J Shelton, Collette M Bloom, and J Kenyatta Cavil, *The significance of hbcus to the production of stem graduates: Answering the call.*, Educational Foundations **26** (2012), 33–47.
[14] Laura Perna, Valerie Lundy-Wagner, Noah D Drezner, Marybeth Gasman, Susan Yoon, Enakshi Bose, and Shannon Gary, *The contribution of hbcus to the preparation of african american women for stem careers: A case study*, Research in Higher Education **50** (2009), no. 1, 1–23.
[15] Terry Platt, Eugene Barber, Antoine Yoshinaka, and Vicki Roth, *An innovative selection and training program for problem-based learning (pbl) workshop leaders in biochemistry*, Biochemistry and Molecular Biology Education **31** (2003), no. 2, 132–136.
[16] Frank C Richardson and Richard M Suinn, *The mathematics anxiety rating scale: psychometric data.*, Journal of counseling Psychology **19** (1972), no. 6, 551.
[17] Ellen L. Usher and Frank Pajares, *Self-efficacy for self-regulated learning: a validation study*, Educ. Psychol. Meas. **68** (2008), no. 3, 443–463, DOI 10.1177/0013164407308475. MR2432235
[18] Carl C Wamser, *Peer-led team learning in organic chemistry: Effects on student performance, success, and persistence in the course*, Journal of Chemical Education **83** (2006), no. 10, 1562.
[19] Linda Reichwein Zientek, Carlton J Fong, and Julie M Phelps, *Sources of self-efficacy of community college students enrolled in developmental mathematics*, Journal of Further and Higher Education **43** (2019), no. 2, 183–200.

Mathematics Department, University of the Virgin Islands, St. Thomas, Virgin Islands 00802
Email address: nmonros@uvi.edu

Innovative Learning Center, LLC Division of Research and Evaluation, Atlanta, Georgia 30309
Email address: ablackmon@ilearningconcepts.com

Office of the Provost, University of the Virgin Islands, St. Thomas, Virgin Islands 00802
Email address: cmckayl@uvi.edu

Mathematics Department, University of the Virgin Islands, St. Thomas, Virgin Islands 00802
Email address: rstolz@uvi.edu

College of Science and Mathematics, University of the Virgin Islands, St. Thomas, Virgin Islands 00802
Email address: sromano@uvi.edu

Contemporary Mathematics
Volume **759**, 2020
https://doi.org/10.1090/conm/759/15269

A discreteness algorithm for 4-punctured sphere groups

Caleb Ashley

In memoriam: This paper is dedicated to the memory of my beloved friend and mentor, Ralph Brooks Turner, Ph.d. 'Last of the free spirits.'

ABSTRACT. Let Γ be a subgroup of $\mathrm{PSL}(2,\mathbf{R})$ generated by three parabolic transformations. The main goal of this paper is to present an algorithm to determine whether or not Γ is discrete. Historically discreteness algorithms have been considered within several broader mathematical paradigms: the *discreteness problem*, the construction and deformation of *hyperbolic structures* on surfaces and notions of *automata* for groups. Each of these approaches yield equivalent results. The second goal of this paper is to give an exposition of the basic ideas needed to interpret these equivalences, emphasizing related works and future directions of inquiry.

1. Introduction

Let Γ be a subgroup of $\mathrm{PSL}(2,\mathbf{R})$ generated by three parabolic transformations. The main goal of this paper is to present an algorithm to determine whether or not Γ is discrete. Historically discreteness algorithms have been considered within several broader mathematical paradigms: the *discreteness problem*, the construction and deformation of *hyperbolic structures* on surfaces and notions of *automata* for groups. Each of these approaches yield equivalent results. The second goal of this paper is to give an exposition of the basic ideas needed to interpret these equivalences, emphasizing related works and future directions of inquiry.

Discreteness is central to the theory of *geometric structures on manifolds.* Given a topological manifold M, by a geometric structure on M, or a (G,X)*-structure* for short, one understands M as being *locally modeled* by the Lie group G on the locally homogeneous space X associated to G [**T1**] [**Gw4**]. For example if G is $\mathrm{Isom}(X)$, the *isometry group* of X, the action of the Lie group G on X preserves a *Riemannian metric* on X, and hence on M. The construction of (global) *symmetric spaces* $X = G/K$ where G is a classical Lie group and K is a maximal compact subgroup of G demonstrates globally symmetric spaces are examples of locally homogenous spaces [**WM**]. Specific examples prominently featured in our discussion here are the *hyperbolic plane* $\mathbb{H} \simeq \mathrm{SL}(2,\mathbf{R})/\mathrm{SO}(2,\mathbf{R})$ and *hyperbolic 3-space* $\mathbb{H}^3 \simeq \mathrm{SL}(2,\mathbf{C})/\mathrm{SU}(2,\mathbf{C})$, along with their respective *isometry groups* $\mathrm{PSL}(2,\mathbf{R})$ and $\mathrm{PSL}(2,\mathbf{C})$.

2020 *Mathematics Subject Classification.* Primary 57Mxx.

Key words and phrases. The discreteness problem, discrete subgroups of $\mathrm{PSL}(2,\mathbf{R})$, hyperbolic geometry, $\mathrm{SL}(2,\mathbf{C})$-character varieties, (G,X)-structures.

EXAMPLE 1.1. Let $\Gamma < \mathrm{PSL}(2, \mathbf{R})$, be a finitely generated non-elementary subgroup. If Γ acts on $\mathbb{H}$ properly discontinuously and freely, then $\Gamma \backslash \mathbb{H}$ is biholomorphic to a Riemann surface which admits a complete hyperbolic metric. Notice the use here of the Uniformization theorem, identifying the *conformal structure* with *hyperbolic structure* [**B**] [**K**]. Γ is an example of a *Fuchsian* group.

EXAMPLE 1.2. Let $\Gamma < \mathrm{PSL}(2, \mathbf{C})$ be a finitely generated non-elementary subgroup. If Γ acts on $\mathbb{H}^3$ properly discontinuously and freely, then $\Gamma \backslash \mathbb{H}^3$ is homeomorphic to the interior of a compact 3-manifold which admits a complete hyperbolic metric. Notice the analogous uniformization theorem here is W. Thurston's classification of hyperbolic 3-manifolds, in particular the version of *Geometrization Conjecture*[1] proved by Thurston [**T2**] [**Md**]. Γ is an example of a *Kleinian* group.

For a fixed locally homogeneous space X with isometry group $\mathrm{Isom}(X)$, the discreteness problem may be described as the creation of a *procedure* which certifies whether or not a given subgroup Γ of $\mathrm{Isom}(X)$ is discrete. A procedure which *halts* in a finite number of steps is called an *algorithm*. Discreteness algorithms were originally pursued in the context of Examples 1.1 and 1.2 [**S1**] [**R**] [**GM**]. Parametrizing $(G, X)-$structures admitted by some manifold M leads to a very rich theory [**BG**] [**B3**] [**C2**] [**Gw4**] [**Gw5**] [**L1**] [**Md**] [**M2**] [**Pa**] [**T1**] [**T2**] [**Wt**]. Discreteness — within *any* programme to investigate the manner in which a given manifold admits geometric structures and the interplay between these structures — must be understood as central precisely by its exhibition of geometry via *symmetry*. Herein we are primarily concerned with the particular case of *hyperbolic structures* on surfaces. In particular, we understand Γ as defined in *Theorem* 1.4 below as answering the following question: Does Γ determine a *complete hyperbolic structure* on the *four-punctured sphere* $S_{0,4}$?

One strategy for understanding subgroups $\Gamma < \mathrm{PSL}(2, \mathbf{R})$ is to construct good *fundamental domains* for the action of Γ on $\mathbb{H}$. *Reduction theory* (the theory of constructing "good" fundamental domains for group actions) has a long history, from Gauss's work "Reduction Theory of Quadratic Forms," to Borel–Harish-Chandra's "Arithmetic subgroups of algebraic groups." As an example of this kind of phenomenon, it is a theorem of Siegel that $\Gamma \backslash \mathbb{H}$ has finite area if and only if Γ is *finitely generated.* Whereby the existence of a finite sided *fundamental domain* $\triangle$ ($\triangle$ is a convex polygon possibly with vertices on the boundary of $\mathbb{H}$) for the Γ action on $\mathbb{H}$ corresponds to combinatorial-group-theoretic information, namely Γ is *finitely generated.* However the analogue of this fact is not necessarily true for Γ Kleinian. There are finitely generated Kleinian groups which are not *geometrically finite* [**C2**].

Given an arbitrary *non-elementary* subgroup $\Gamma < \mathrm{PSL}(2, \mathbf{R})$, a necessary condition for discreteness is *Jørgensen's inequality* [**J**]. A consequence of Jørgensen's inequality is the following (rather intractable example of a) sufficient condition for discreteness: Γ is discrete only if every *non-elementary rank 2* subgroup is discrete. Rank 2 means Γ is generated by two elements, $\Gamma = \langle A, B \rangle$ for $A, B \in \mathrm{PSL}(2, \mathbf{R})$. The *Poincaré polygon theorem* is an example of a more *effective* sufficient condition, more effective in the sense that it holds for polygons $\triangle$ embedded in the hyperbolic plane $\mathbb{H}$ with rational endpoints [**M1**]. However, necessary and sufficient criteria for discreteness *can not* exist, via the following work of Gilman-Maskit.

[1] Thurston's revolutionary work first suggested an analogous theorem to Uniformization might be true in 3-dimensions [**T2**].

THEOREM 1.3. *[Gilman-Maskit] For* $\Gamma < \mathrm{PSL}(2, \mathbf{R})$, *a non-elementary rank 2 subgroup, this discreteness problem was settled by J. Gilman & B. Maskit* [**GM**]. *The solution to the discreteness problem is an algorithm, we will denote it by the* $GM-$*algorithm. Via analysis of the complexity of this algorithm, no necessary and sufficient condition for discreteness exists* [**GM**] [**Gl2**].

The $GM-$algorithm couples algebraic and geometric conceptual procedures, each one informing the other. Ultimately, the method of proof utilizes the intrinsic geometry of the hyperbolic plane to determine if a *fundamental domain* $\triangle$, can be built for the Γ action on $\mathbb{H}$. The following first appeared in [**A**].

THEOREM 1.4. *[A.] Let* A, B *and* C *be parabolic elements of* $\mathrm{PSL}(2, \mathbf{R})$. *For* $\Gamma = \langle A, B, C\rangle$ *non-elementary free subgroup, a procedure to determine whether or not* Γ *is discrete exists. Furthermore, the procedure is an algorithm if* Γ *has no elliptic elements, we will denote as the* $4PS-$algorithm, *since* Γ *corresponds to a four-punctured sphere group.*

OUTLINE: In Section 2 we present fundamental ideas of hyperbolic structures on surfaces. In Section 3 we summarize the $GM-$algorithm and notions of automata for groups. In Section 4 we give a self contained proof of Theorem 1.4. In Section 5 we conclude with brief interpretations and indications of future directions of inquiry.

2. Background

The purpose of this section is to habilitate the $4PS-$algorithm within the mathematical theory in which it lives. We begin with examples of constructions of Riemann surfaces, and proceed with deformations of geometric structures on surfaces, focusing on hyperbolic structures in the paradigm of locally homogenous structures.

DEFINITION 2.1. A *Riemann surface* S is a 1-dimensional complex manifold.

As a consequence of Definition 2.1, Riemann surfaces are orientable, triangulizable smooth surfaces on which a certain collection complex valued continuous functions are designated as *holomorphic*. One way of defining such a *conformal structure* on a given oriented smooth surface is to construct (via Gauss) a Riemannian metric and define *local coordinate charts* as conformal homeomorphisms to $\mathbf{C}$ [**B3**].

DEFINITION 2.2. A Riemann surface S is called *hyperbolic*, if it admits a *Riemannian metric* $ds_{\mathbb{H}}$ of constant curvature negative curvature, where the boundary of S (if nonempty) is totally geodesic [**FM**].

2.1. Construction of Riemann surfaces. We desire to distinguish between Riemann surfaces S and the underlying topological surface Σ. We will denote *closed* (compact, no boundary) Riemann surfaces of *genus* g by S_g and Riemann surfaces of *finite type* (genus g, n *marked points*) by $S_{g,n}$; similarly Σ_g and $\Sigma_{g,n}$ (genus g, n punctures) denote the underlying topological surfaces of S_g and $S_{g,n}$, respectively. The fundamental group of any surface will be denoted by π_1.

EXAMPLE 2.3. (Algebraic Construction) Riemann surfaces[2] were originally described as maximal domains on which holomorphic functions (themselves multi-valued via the process of *analytic continuation* on domains Ω of $\mathbf{C}$) become single-valued. So described, Riemann surfaces are manifestly *algebraic*, namely as n-sheeted *branched coverings* of the complex projective line, $\mathbf{C}P^1$. The details of this construction yield a canonical method of *compactification*, thus there is a natural correspondence between *irreducible smooth projective algebraic curves* defined over $\mathbf{C}$ and *compact* Riemann surfaces [**D**] [**FsKa**] [**FrKl**].

EXAMPLE 2.4. (Arithmetic Construction) A natural question regarding the algebraic construction is: when does a Riemann surface correspond to an irreducible algebraic curve defined over $\overline{\mathbf{Q}}$? Examples invovle *the Modular group* PSL$(2, \mathbf{Z})$, *congruence subgroups*, or any *arithmetic subgroup* of PSL$(2, \mathbf{R})$ [**FrKl**] [**Rd**].

EXAMPLE 2.5. (Topological Construction) The topological classification of surfaces gives a combinatorial representation of π_1, as a $(4g + n)$−gon with $2g$ side-pairing identifications:

$$\pi_1 =< a_1, b_1, \ldots, a_g, b_g, c_1, \ldots, c_n \; || [a_1, b_1] \cdots [a_g, b_g] \cdot c_1 \cdots c_n = 1 > . \tag{2.1}$$

$[a_i, b_i] = a_i \cdot b_i \cdot {a_i}^{-1} \cdot {b_i}^{-1}$ denotes the *commutator* of the side-pairing generators a_i and b_i and the c_i are boundary generators [**D**]. The *Euler characteristic*, denoted $\chi(\pi_1)$ is a topological invariant and is given by $(2 - 2g) - n$. It is a classical result, that if $n \geq 1$ then π_1 is isomorphic to a *free group* of rank N, denoted $\mathbf{F}_N$, where $N = 2g + n + 1$. For example, with Γ as in Theorem 1.4, $\pi_1 \simeq \mathbf{F}_3$, a free product of three *infinite-cyclic* groups [**St**].

EXAMPLE 2.6. [Uniformization Theorem: Poincaré, Kobe, Klein] If X is a simply connected Riemann surface, then X is isomorphic to $\mathbf{C}P^1$, $\mathbf{C}$ or $\mathbb{H}$ [**Spr**].

The isomorphism above is with respect to the Riemann surface structure. I.e. the *uniformizing* isomorphism is a *diffeomorphism* that respects the holomorphic structure, i.e. it is a *holomorphic isomorphism* [**D**]. Since holomorphic maps are conformal, near every point of S (locally) the metric alluded to in the remarks after Definition 2.1 can be written as $ds = \lambda(z)|dz|$, where z is the "uniformizing" parameter, and an appropriate choice of *conformal factor* $\lambda(z)$, can be made so κ is -1.

An equivalent statement of uniformization is: every connected Riemann surface S is biholomorphic to $\Gamma \backslash X$, where $X = \mathbf{C}P^1, \mathbf{C}$, or $\mathbb{H}$ and Γ acts on X *properly discontinuously*, *freely*, and by *biholomorphisms* [**B**]. This implies:

(1) π_1 acts by isometries on X,
(2) π_1 is isomorphic to Γ, where Γ is a discrete subgroup of Isom(X).

The π_1 action is by isometries, since the conformal automorphisms of X are precisely the isometries of X, i.e. Aut(X) = Isom(X). —This is often described as "a miracle" of low-dimensional topology! Uniformization is a powerful conceptual bridge, providing a correspondence between *conformal structures* and *metric structures* on a surface. For example, thinking of a Riemann surface $S_{g,n}$ with negative Euler characteristic as coming from a uniformization identifies π_1 isomorphically with a discrete subgroup Γ of PSL$(2, \mathbf{R})$. Indeed if $\chi(\pi_1) < 0$, the *Gauss-Bonnet*

[2]Riemann surfaces first appeared in the Ph.D thesis of Bernhard Riemann, [1851] "Theory of functions of one complex variable."

theorem implies that Σ is a Riemann surface which admits a metric of constant negative curvature κ, hence its universal cover is $X = \mathbb{H}$, and $\text{Isom}(X) = \text{PSL}(2, \mathbf{R})$.

REMARK 2.7. It is worthwhile to note that uniformization is a deep analytic result and is by no means a straightforward constructive process. Also, Examples 2.5 and 2.6 imply that the Riemann surface detected by Theorem 1.4 must be the four-punctured sphere, $\Sigma_{0,4}$. Since $\Gamma \simeq \mathbf{F}_3$ and Γ acting freely on $\mathbb{H}$ as in Theorem 1.4 imply $\Gamma\backslash\mathbb{H} \simeq S$ has four *ends* which are *cusps*. I.e. $\Gamma\backslash\mathbb{H} \simeq S$ has four ends which are complete and have finite area.

The following dynamical construction is a one of the first methods of building examples of Fuchsian and Kleinian groups [**B**] [**CT**].

EXAMPLE 2.8. (Ping-Pong Lemma) Let $A, B \in \text{PSL}(2, \mathbf{R})$ and let ${U_A}^-, {U_A}^+$, ${U_B}^-, {U_B}^+$ be pairwise disjoint *half-spaces* $\subset \mathbb{H}$ such that ${U_A}^+$ is *pairwise disjoint* with ${U_B}^-$ and ${U_A}^-$ is *pairwise disjoint* with ${U_B}^+$. If $A \cdot (\mathbb{H}\backslash\{\overline{U_A^-}\}) = {U_A}^+$ and $B \cdot (\mathbb{H}\backslash\{\overline{U_B^-}\}) = {U_B}^+$ then all of the following are true:

(1) $\langle A, B\rangle = \Gamma < \text{PSL}(2, \mathbf{R})$ is freely generated by A and B,
(2) Γ is discrete,
(3) Γ acts freely on $\mathbb{H}$,
(4) $\forall\gamma \in \Gamma\backslash\{\text{Id}\}$, γ is hyperbolic,
(5) $\mathbb{H}\backslash\{\{\overline{U_A^-}\}) \cup \{U_A^+\}) \cup \{\overline{U_B^-}\} \cup \{U_B^+\}\}$ is a fundamental domain for the Γ action on $\mathbb{H}$.

The following geometric construction is a sufficient condition for discreteness and a main part of the conceptual framework of the $GM-$algorithm [**M1**] [**B**]

EXAMPLE 2.9 (Poincaré Polygon Theorem). For all $d \geq 5$, there exists a *regular* right-angled $d-$gon $\triangle$ embedded in $\mathbb{H}$, so that Γ (the *double-cover* of the subgroup generated by reflections along the sides of $\triangle$) is a discrete subgroup of $\text{PSL}(2, \mathbf{R})$ is discrete. Furthermore $\triangle$ is a fundamental domain for the Γ action on $\mathbb{H}$ and $S \simeq \Gamma\backslash\mathbb{H}$ is closed or finite area hyperbolic surface. [Note that the Riemann surface constructed in this manner may be a hyperbolic *orbifold*. I.e. $S \simeq \Gamma\backslash\mathbb{H}$ may be an *incomplete* hyperbolic structure, meaning the hyperbolic metric may have *conical singularities*.]

2.2. Hyperbolic geometry. The intrinsic geometry of the *hyperbolic plane*[3] may be described via tools of Riemannian geometry as a (real) 2-dimensional *homogeneous* space which admits a metric of constant negative *curvature* [**B**] [**CT**] [**K**] [**Mi1**] [**Md**]. A *model* for the hyperbolic plane is a pair (X, ds_X), where X denotes the underlying set and ds_X denotes the metric. The *Upper Half Plane* model of the hyperbolic plane is: $\left(\mathbb{H} = \{x + iy \in \mathbf{C}|\ y > 0\}\ ,\ ds_{\mathbb{H}} = \frac{dx^2+dy^2}{y^2}\right)$.

(2.2)
$$\text{PSL}(2, \mathbf{R}) := \text{SL}(2, \mathbf{R})/\{\pm\mathbb{I}\} = \{\begin{bmatrix} a & b \\ c & d \end{bmatrix} : a, b, c, d \in \mathbf{R}, ad - bc = 1\}/\{\pm\mathbb{I}\}.$$

$\text{PSL}(2, \mathbf{R})$ acts on $\mathbb{H}$ via *Möbius transformations*. That is, for any $z \in \mathbb{H}$, and for any $A \in \text{PSL}(2, \mathbf{R})$ (equivalently $-A$), $A \cdot z \mapsto \frac{az+b}{cz+d}$. The metric $ds_{\mathbb{H}}$ induces a notion of hyperbolic distance denoted $d_{\mathbb{H}}(z, w)$ between any two points z, w in $\mathbb{H}$.

[3]There are several models for the hyperbolic plane. Each model has particular advantages and disadvantages for understanding phenomena in the hyperbolic plane [**B**] [**K**].

By joining z to w with a smooth curve γ in $\mathbb{H}$, the *hyperbolic length* denoted $\ell_{\mathbb{H}}(\gamma)$ is given by $\int_\gamma |ds_{\mathbb{H}}|$. Finally, the hyperbolic distance $d_{\mathbb{H}}(z,w)$ is defined as $\inf_\gamma \ell_{\mathbb{H}}(\gamma)$.

DEFINITION 2.10. A mapping $f : \mathbb{H} \to \mathbb{H}$ is an *isometry* if it preserves the hyperbolic distance, that is $d_{\mathbb{H}}(f(z), f(w)) = d_{\mathbb{H}}(z,w)$, for all $z, w \in \mathbb{H}$.

Viz à viz the overarching principle of symmetry, isometries are objects of extreme utility, as are *geodesics.* For any two points x, y not on the boundary, there is a unique geodesic between them. $\mathbb{H}$ is therefore a *geodesic space.* The boundary of $\mathbb{H}$, denoted by $\partial_\infty\mathbb{H}$, is the *real projective line* $\mathbf{R}P^1 := \mathbf{R} \cup \{\infty\}$. $\mathrm{PSL}(2,\mathbf{R}) = \mathrm{Isom}^+(\mathbb{H})$, the *orientation preserving isometries* of $\mathbb{H}$. This follows from the fact that $\mathrm{PSL}(2,\mathbf{C})$ is the group of conformal automorphisms of the complex projective line $\mathbf{C}P^1$, where the action is also by Möbius transformations but the coefficients are restricted from $\mathbf{C}$ to $\mathbf{R}$. The restriction of the action from $\mathrm{PSL}(2,\mathbf{C})$ to $\mathrm{PSL}(2,\mathbf{R})$ is characterized by fixing a $\mathbf{R}P^1$ inside $\mathbf{C}P^1$, equivalently $\mathrm{PSL}(2,\mathbf{R})$ preserves a $\mathbb{H} \subset \mathbf{C}P^1$. [This gives an alternative description of the hyperbolic metric $ds_{\mathbb{H}}$ as a conformally Euclidean metric.] Furthermore, identifying $\mathbf{R}^2$ with $\mathbf{C}$ and using the geometry of complex numbers, along with the fact that Möbius transformations are generated by an even number of compositions of reflections in circles and lines, all isometries of $\mathbb{H}$ can be understood to be generated by (possibly finite compositions of) the following: *reflection* about $y-$axis $(x,y) \mapsto (-x,y)$, *horizontal translation* $(x,y) \mapsto (x+c,y)$, *scaling* or *dilatation* $(x,y) \mapsto (cx,cy)$, *inversion in the unit circle* $(x,y) \mapsto (\frac{x}{x^2+y^2}, \frac{y}{x^2+y^2})$ [**Md**].

DEFINITION 2.11. Submanifolds of (real) dimension one which minimize distance are called *geodesics.*

Via horizontal shifting and scaling, all geodesics of $\mathbb{H}$ are characterized as either vertical lines or circular arcs that intersect $\mathbf{R}$ orthogonally. Similarly isometries may be classified geometrically by their *fixed points.* $A \in \mathrm{PSL}(2,\mathbf{R})$ is called: *elliptic* if A has one fixed point in $\mathbb{H}$, *hyperbolic* if A has two fixed points on $\partial_\infty\mathbb{H}$, *parabolic* if A has *exactly* one fixed point on $\partial_\infty\mathbb{H}$. Irrespective of the model of the hyperbolic plane, geodesics will be fixed as sets (not point-wise but globally) of *involutions.* It is worthwhile to observe that hyperbolic isometries may also be characterized by open half-panes, as in the ping-pong construction of Example 2.8. In particular, *disjoint pairwise orthogonal* geodesics determine disjoint *open half-spaces* which give a fundamental domain for the action of a hyperbolic isometry. Alternatively isometries are classified algebraically. $A \in \mathrm{PSL}(2,\mathbf{R})$ as in Equation 2.2 is elliptic, hyperbolic, or parabolic if $|\mathrm{Tr}A| < 2, |\mathrm{Tr}A| > 2$, or $|\mathrm{Tr}A| = 2$, respectively where $|\mathrm{Tr}| := |a+d|$ denotes the absolute value of the trace of A. The algebraic characterizations of geodesics are independent of model of the hyperbolic plane [**B**] [**Md**].

The topology induced on $\mathbb{H}$ by either the hyperbolic or Euclidean metric is the same. Therefore the action of any $\Gamma < \mathrm{PSL}(2,\mathbf{R})$ *extends* to the *Euclidian closure* of $\mathbb{H}$, denoted $\overline{\mathbb{H}} := \mathbf{R} \cup \{\infty\}$. For $z_0 \in \mathbb{H}$, the set of all possible *limit points* of Γ-orbits $\Gamma \cdot z_0$, is called the *limit set* of Γ and is denoted by $\Lambda(\Gamma)$. It can be shown that $\Lambda(\Gamma)$ does not depend on the point z_0. If Γ is discrete then $\Lambda(\Gamma)$ is necessarily a subset of $\overline{\mathbb{H}}$ [**B**].

DEFINITION 2.12. A discrete subgroup Γ of $\mathrm{PSL}(2,\mathbf{R})$ is called *elementary* if it has a finite orbit in $\Lambda(\Gamma)$. Otherwise, Γ is called *non-elementary.*

A non-elementary subgroup Γ of $\mathrm{PSL}(2, \mathbf{R})$ must have one hyperbolic element. In fact, any such Γ has infinitely many such elements, no two of which have a common fixed point. Many algebraic, geometric, and dynamic consequences can be derived if a group is non-elementary [**B**] [**K**].

DEFINITION 2.13. Discrete non-elementary subgroups of $\mathrm{PSL}(2, \mathbf{R})$ are called Fuchsian. Discrete non-elementary subgroups of $\mathrm{PSL}(2, \mathbf{C})$ are called Kleinian.

The following proposition is a necessary condition for discreteness and is a main part of the conceptual framework of $GM-$algorithm.

PROP 2.14. *(Jørgensen's inequality)* [**J**] *If $\langle A, B\rangle$ is a non-elementary Kleinian group then:*

$$|\mathrm{Tr}(A)^2 - 4| + |\mathrm{Tr}([A, B]) - 2| \geq 1.$$

2.3. Dynamics of Fuchsian groups. The *covering theory* implicit in the examples in Section 2.1 leads to a fundamental dynamical question: When and where does a group act *properly discontinuously*?

DEFINITION 2.15. A group Γ acts *properly discontinuously* on a topological space X if for every compact set $K \subseteq X$, $\{\gamma \in \Gamma | \;\; \gamma \cdot K \cap K \neq \emptyset\}$ is finite.

DEFINITION 2.16. For Γ and X as in Definition 2.15, a region $\Omega \subseteq X$ where Γ acts properly discontinuously is called a *domain of discontinuity*, denoted by $\Omega(\Gamma)$.

Let M be a topological manifold and Γ be a group, Γ is said to act on M by *orientation preserving homeomorphism* if and only if there exists a *non-trivial* homomorphism $\rho : \Gamma \to \mathrm{Homeo}^+(M)$. Via the *compact-open* topology, $\mathrm{Homeo}^+(M)$ is a topological group [**CT**]. There is a natural bijective correspondence between actions of Γ on M and representations of Γ into $\mathrm{Sym}(X)$ [**Sh**]. An action is said to be *effective* if it corresponds to a *faithful* representation [**Sh**]. Along with a *fixed-point* theorem which describes canonical neighborhoods of *torsion* points, properly discontinuous actions give the following desired quotient construction.

PROP 2.17. *If X is a Hausdorff, locally compact topological space and if Γ acts on X by homeomorphisms and Γ acts properly discontinuously and freely on X, then the quotient space $\Gamma\backslash X$ is a Hausdorff topological surface via the quotient topology* [**CT**].

We may characterize discreteness with the following equivalencies: a finitely generated subgroup Γ of $\mathrm{PSL}(2, \mathbf{R})$ acts properly discontinuously on $\mathbb{H}$ if and only if Γ is a discrete subgroup of $\mathrm{PSL}(2, \mathbf{R})$ if and only if there exists a finite-sided convex fundamental domain $\triangle$ for the Γ action on $\mathbb{H}$ [**B**] [**K**].

DEFINITION 2.18. Given a topological manifold M and a group Γ acting on M properly discontinuously, a subset $\Omega \subseteq M$ is called a *fundamental domain* if and only if each of the following hold:

(1) $\overline{\Omega} \longrightarrow \Gamma\backslash M$ is surjective. I.e. every Γ-orbit meets $\overline{\Omega}$,
(2) $\mathrm{Int}(\Omega) \longrightarrow \Gamma\backslash M$ is injective. I.e. $\gamma \cdot \mathrm{Int}(\Omega)$ are disjoint for each $\gamma \in \Gamma$,
(3) $\partial\Omega$, the boundary of Ω, defined by $\partial\Omega := \overline{\Omega}\backslash\mathrm{Int}(\Omega)$, is "small." I.e. for any Γ-*invariant measure* on Ω, the measure of the boundary is zero,
(4) (Local Finiteness) $\forall\gamma \in \Gamma, \forall$ compact subsets $K \subset X$, M meets only finitely many translates of $\gamma \cdot \overline{\Omega}$.

DEFINITION 2.19. For any $z_0 \in \mathbb{H}$ only fixed by $\mathrm{Id} \in \Gamma$, $D_\Gamma(z_0)$ the *Dirichlet fundamental domain* is: $D_\Gamma(z_0) = \{z \in \mathbb{H} \mid \ d(z, z_0) \leq d(z, \gamma \cdot z_0),\ \forall \gamma \neq \mathrm{Id} \in \Gamma\}$.

REMARK 2.20. We now emphasize our focus on the paradigm of locally homogenous structures. Propositions 2.21 and 2.21 demonstrate how it is possible to parametrize the limit set $\Lambda(\Gamma)$ of a hyperbolic structure with the boundary of the hyperbolic plane. Example 2.23 demonstrates how to further employ the limit set $\Lambda(\Gamma)$ to parametrize deformations of hyperbolic structures [**Gw4**] [**L1**] [**T1**]. [These should be understood as motivation for the manner in which a Riemann surface S which admits a hyperbolic metric may be advantageously interpreted as a hyperbolic $(G, X)-$structure on Σ. Recall how G and X are described proceeding Example 1.1 in Section 1; these hyperbolic geometric structures are characterized by two pieces of data: $\rho : \pi_1\Sigma \longrightarrow \mathrm{PSL}(2, \mathbf{R})$ and $\mathrm{Dev} : \widetilde{\Sigma} \longrightarrow \mathbb{H}$.]

PROP 2.21. *Given a representation* $\rho : \pi_1\Sigma_g \longrightarrow \mathrm{PSL}(2, \mathbf{R})$ *such that there exists a mapping* $\xi : \partial_\infty(\pi_1\Sigma_g) \longrightarrow \partial_\infty\mathbb{H}$ *continuous, injective, and* $\rho-$*equivariant (I.e.* ξ *intertwines the action of* $\pi_1 S$ *with action of* ρ *on* $\partial_\infty\mathbb{H}$*), then:*

(1) ρ *is injective,*
(2) ρ *is discrete,*
(3) $\rho(\pi_1\Sigma)$ *is cocompact, that is* $\rho(\pi_1\Sigma)$ *has compact quotient.*

As we have defined Γ in Theorem 1.4, $\Sigma_{g,n} \simeq \Gamma\backslash\mathbb{H}$ does not have boundary. Yet, it is worthwhile (regarding *compactifications*) to consider the case of Riemann surfaces $\Sigma^r_{g,n}$, genus g, n punctures and r boundary components, that is r *ends* that are *collar neighborhoods* of closed geodesic boundary elements.

PROP 2.22. *Given a representation* $\rho : \pi_1\Sigma^r_{g,n} \longrightarrow \mathrm{PSL}(2, \mathbf{R})$ *such that there exists a mapping* $\xi : \partial_\infty(\pi_1\Sigma^r_{g,n}) \longrightarrow \partial_\infty\mathbb{H}$ *continuous, injective and* ρ*-equivariant, then:*

(1) ρ *is injective,*
(2) ρ *is discrete,*
(3) $\rho(\gamma)$ *is hyperbolic,* $\forall\gamma \neq \mathrm{Id} \in \Gamma$.

It is a fact that Dev is $\rho-$equivariant and we will see Dev is an isometry which extends to the boundary of the hyperbolic plane. Therefore Dev can be identified with ξ in Propositions 2.21, 2.21. Furthermore, the limit set Λ_Γ is the entire boundary of the hyperbolic plane, $\partial_\infty\mathbb{H}$. [For surfaces with boundary, like $\Sigma^r_{g,n}$ in Proposition 2.22, Λ_Γ will be a *Cantor set* contained in $\partial_\infty\mathbb{H}$. Indeed, $\partial_\infty\pi_1(\Sigma^r_{g,n})$ is *homotopy equivalent* to the *Cayley graph* of $\pi_1(\Sigma^r_{g,n})$]. In both cases, Dev "encodes" the entire representation ρ. For example the *Hausdorff dimension* of Λ_Γ gives a notion of "geometric magnitude" of a surface, and this is one sense in which a representation ρ is encoded by the *boundary map* Dev [**L1**] [**Gw4**].

EXAMPLE 2.23. (*Quasi-Fuchsian* Structures) R. Riley developed one of the first computer programs for detecting discreteness [**R**]. Riley's procedure detected *quasi-Fuchsian* structures $\mathcal{QF}(\Sigma)$. Given a Fuchsian structure on Σ (a discrete and faithful representation $\rho : \pi_1 \longrightarrow \mathrm{PSL}(2, \mathbf{R})$) a quasi-Fuchsian structure can be constructed by "wiggling" the representation ρ a "bounded" amount so that the image of the limit set $\rho(\Lambda_\Gamma)$ is deformed from $\mathbf{R}P^1$ (Fuchsian) to a *Jordan curve* (quasi-Fuchsian). [The precise way to affect this type of *quasi-conformal* deformation is described via *Beltrami differentials* [**Md**].] It is a fact that $\Gamma \simeq \pi_1$

acts properly discontinuously and cocompactly on $\Omega(\Gamma_\rho) := \mathbf{C}P^1 \backslash \{\xi(\partial_\infty \pi_1)\}$. Also $\xi : \partial_\infty \pi_1 \longrightarrow \partial_\infty \mathbb{H}$, so that the domain of discontinuity $\Omega(\Gamma_\rho)$ is homeomorphic to $\mathbb{H} \cup \mathbb{H}^-$, where $\mathbb{H}^-$ denotes the *Lower Half Plane.* This construction gives a uniformization of a quasi-Fuchsian manifold $M_\rho = \Gamma \backslash (\mathbb{H}^3 \cup \Omega(\Gamma_\rho))$ [**M2**] [**Md**].

THEOREM 2.24. *[Riley] Discreteness is decidable for non-elementary finitely generated subgroups of* $\mathrm{SL}(2, \mathbf{R})$ [**R**] [**Kp2**]. *(See Section* 3.2 *for relevant definitions of decidability models and halting sets.)*

2.4. Deformations of hyperbolic structures. The *deformation theory* of geometric structures is naturally involved in the discreteness problem; we have observed via example the correspondence between the classification of Riemann surfaces and discrete subgroups of the automorphism groups of $\mathbf{C}P^1, \mathbf{C}$, or $\mathbb{H}$. A *deformation space* of geometric structures is a space whose points themselves are geometric structures, whereby deformation spaces parametrize geometric structures admissible on a given topological manifold [**Gw4**]. The *Deformation theory* of geometric structures is quite vast. Two sources which emphasize the themes of algebra and analysis at the heart of complex function theory of Riemann surfaces are [**Mfd**]and [**B3**]. For a series of surveys see [**Pa**]. Also see [**C2**] [**CT**] [**FM**][**Gw1**][**Gw2**] [**Gw3**] [**Gw4**] [**Gw5**] [**L1**] [**Md**] [**T1**] [**WW**] [**Wt**]. There are many different constructions of deformation spaces of Riemann surfaces: Mumford's algebraic construction of $\mathcal{M}_g$ (the moduli space of stable algebraic curves), Teichmüller's complex-analytic construction of $\mathcal{T}(S_g)$ (the space of conformal structures on S_g), Hitchin's complex gauge-theoretic construction of $\mathfrak{M}_{n,g}$ (the moduli space of rank$-n$ stable Higgs bundles on S_g.) We will briefly describe Fenchel-Nielsen's real-analytic construction of the *Fricke space* $\mathcal{F}(\Sigma_g)$, the space of *complete marked hyperbolic structures* on Σ_g. This construction emphasizes uniformization, and is based on work of Fricke-Klein which develops the deformation theory of hyperbolic structures on a surface in terms of the space of representations of its fundamental group π_1 into $\mathrm{SL}(2, \mathbf{C})$ [**FrKl**] [**Gw1**].

DEFINITION 2.25. A *marked* hyperbolic structure is a pair (S_1, ϕ_1), where S and S_1 are Riemann surfaces and $\phi_1 : S \longrightarrow S_1$ is a diffeomorphism such that if there exists an isometry $I : S_1 \longrightarrow S_2$, then $\phi_1 \simeq I \circ \phi_2$ up to *homotopy* [**FM**].

A marking defines an equivalence relation on hyperbolic structures [**FM**]. Also, since any two Riemann surfaces of constant curvature -1 are locally isometric and since via uniformization any local isometry from a connected subdomain of $\mathbb{H}$ *uniquely* extends to an isometry on all of $\mathbb{H}$, a marked structure determines two pieces of data: a map which globalizes the coordinate charts of S, $\mathrm{Dev} : \widetilde{S} \longrightarrow X$ and a representation which globalizes the coordinate changes of S, $\rho : \pi_1 \longrightarrow G$. Dev is called *developing map*, ρ is called *holonomy representation* and X is the homogeneous space of G. This is precisely the data of a (G, X)-structure on Σ [**Gw4**].

DEFINITION 2.26. The pair (Dev, ρ) is called a *developing pair* and determines a (G, X)-structure on Σ.

DEFINITION 2.27. The $\mathrm{SL}(2, \mathbf{R})$-*representation variety*, denoted $\mathcal{R}(\pi_1)$, is given by: $\mathcal{R}(\pi_1) = \mathrm{Hom}(\pi_1, \mathrm{SL}(2, \mathbf{R}))/\mathrm{SL}(2, \mathbf{R})$, where $\mathrm{SL}(2, \mathbf{R})$ acts by conjugation.

It is possible to construct $\mathcal{D}_{(G,X)}(\Sigma)$, the deformation space of all (G, X)-structures on Σ [**Gw4**]. The points of such deformation spaces are isomorphism classes of (G, X)-structures on Σ. When $G = \mathrm{PSL}(2, \mathbf{C})$ these deformation spaces are called *Fricke space* $\mathcal{F}(\Sigma_g)$. Furthermore, Mod *the mapping class group* of Σ, whose elements maybe be described as *isotopy* classes of elements of $\mathrm{Homeo}^+(\Sigma)$ which fix the boundary point-wise (if non-empty) and preserve the set of punctures, acts on $\mathcal{D}_{(G,X)}(\Sigma)$ by changing the marking on the underlying surface [**FM**].

THEOREM 2.28. *(Fricke) Let G be* $\mathrm{SL}(2, \mathbf{C})$, Mod *acts properly discontinuously but not necessarily freely on* $\mathcal{F}(S_g)$.

The holonomy representation defines a mapping from the deformation space of (G, X)-structures to the G-*representation variety*:

$$\rho : \mathcal{D}_{(G,X)}(\Sigma) \longrightarrow \mathrm{Hom}(\pi_1, G)/G. \tag{2.3}$$

It can be shown that ρ is Mod-equivariant [**Gw4**] [**L1**] [**T1**].

THEOREM 2.29. *(Thurston) In Equation* 2.3, *ρ is a local homeomorphism. Furthermore if Σ be Σ_g, then ρ is an embedding. (This is a theorem of A. Weil when $G =$* $\mathrm{PSL}(2, \mathbf{R})$, *and is implicit in the work of C. Ehrasman)* [**Gw4**].

REMARK 2.30. In $4PS-$algorithm, $G = \mathrm{PSL}(2, \mathbf{R})$, $\pi_1 \simeq \mathbf{F}_3$, and π_1 acts freely on $\mathbb{H}$. This deformation space is the Fricke space $\mathcal{F}(\Sigma)$ contains equivalence classes of *complete marked* hyperbolic structures on $\Sigma_{0,4}$.

For closed surfaces the uniformization theorem can be reinterpreted in this context as identifying $\mathcal{F}(\Sigma_g)$ with $\mathcal{T}(S_g)$ [**Gw1**]. In fact, the holonomy representation ρ embeds $\mathcal{F}(\Sigma_g)$ as a *connected component* of $\mathrm{Hom}(\pi_1, \mathrm{PSL}(2, \mathbf{R}))/\mathrm{PSL}(2, \mathbf{R})$. It is a theorem of Goldman that representations ρ with maximal *Euler class* $e(\rho)$, (maximal with respect to the *Milnor-Wood inequality*: $|e(\rho)| \leq |\chi(\Sigma)|$) are discrete and faithful and correspond to an entire component of $\mathcal{R}(\pi_1)$ [**Gw2**]. Note that $e(\rho)$ may be understood as a generalization of *rotation number* of a homeomorphism. [Just as the unit tangent bundle on a surface corresponds to the space of directions, the fiber-wise restriction a Riemannian metric on a vector bundle endows each fiber with an *angular form* ψ. In particular, $e(\rho)$ is defined as the *differential* of the angular form $d\psi$ in each fiber of a *sphere-bundle* on S [**Mi2**] [**Gw2**].] Moreover Goldman shows every integer suggested by the Milnor-Wood occurs (there are $4g - 3$ components of $\mathcal{R}(\pi_1)$ for S_g), and that the Mod action is not properly discontinuous on the non-maximal components of $\mathcal{R}(\pi_1)$ [**Gw2**]. The behavior of non-maximal components is more mysterious dynamically, but by no means less interesting, they contain points which correspond to *singular* hyperbolic structures [**Gw3**].

The most natural way to understand the topology of $\mathcal{F}(\Sigma_g)$ is via Fenchel-Nielsen coordinates, which are *length-twist* coordinates on pairs of *hyperbolic pants*. These homeomorphically identify $\mathcal{F}(\Sigma_g)$ with $\mathbf{R}^{6g-6}$. In particular, the deformation space of complete hyperbolic structures on Σ_g has real dimension $dim_{\mathbf{R}} = 6g - 6$.

Certain *algebro-geometric-analytic* objects called *character varieties*, denoted here by $\mathfrak{X}_\Gamma(G)$, parametrize the deformation spaces of hyperbolic structures on Σ [**Gw1**] [**ABL**]. Character varieties are obtained from representation varieties by forming an appropriate *geometric invariant theory* quotient. Precisely, we make the following identification of orbits:

$$\mathfrak{X}_\Gamma(G) := \mathrm{Hom}(\Gamma, G)//G, \text{where } [\rho_1] \sim [\rho_2] \text{ if and only if } \overline{[\rho_1]} \cap \overline{[\rho_2]} \neq \emptyset.$$

A theorem of Procesi is $\mathfrak{X}_\Gamma(G)$ is generated by trace functions, $|\mathrm{Tr}|$ [**ABL**] [**Gw1**].

THEOREM 2.31. *[Goldman] In* [**Gw1**], *the* $\mathrm{SL}(2,\mathbf{C})$*-character variety of a rank 2 free* $\mathbf{F}_2$ *is identified with* $\mathbf{C}^3$*; furthermore, by relating the* FN*-coordinates of the three-holed sphere* $\Sigma_{0,3}$ *to trace parameters corresponding to the boundary components, the Fricke space of* $\Sigma_{0,3}$ *is identified with* $(-\infty,-2]^3$. *Goldman also gives a geometric description of the irreducible* $\mathrm{SL}(2,\mathbf{C})$*-representations of* $\mathbf{F}_2$ *as "mildly degenerate" hexagons, from which he builds a Coxeter extension of* $\mathbf{F}_2$*, via reflection in the common perpendiculars of the axes; ultimately describing the Fricke space of the one-holed torus* $\Sigma_{1,1}$. *See* [**Gw1**] *for details, including descriptions of the the character varieties of* $\Sigma_{0,4}$ *and* $\Sigma_{1,2}$.

THEOREM 2.32. *[Maloni-Palesi-Tan] In* [**MPT**], *using Morse theoretic techniques the* Mod *action on relative* $\mathrm{SL}(2,\mathbf{C})$*-character varieties of* $\Sigma_{0,4}$ *is described. In particular, the existence of a non-empty domain of discontinuity for* $\mathrm{PSL}(2,\mathbf{R})$ *representations of 4-punctured sphere groups is demonstrated. Whereby regions of* $\mathrm{PSL}(2,\mathbf{R})$*-character varieties of* $\Sigma_{0,4}$ *whose points correspond to singular hyperbolic metrics on* $\Sigma_{0,4}$ *are parametrized.*

3. Automata

Now we turn to notions of *decidability* and *complexity* of algorithms for groups. The comparative analysis and complexity of algorithms requires making a lot of choices [**BBS**] [**E**] [**Gl1**] [**Gl2**] [**S1**] [**We**] [**Kp1**]. The choices may seem naive, but they are actually quite subtle. For example, what does one mean by algorithm? Or what does one mean by decidability? — The complexity analysis of algorithms is more qualitative but choice remains. For example, complexity should measure what? We follow [**GM**] [**Gl2**], taking an *algorithm* to be a procedure which stops in finite time on *all* input values, as apposed to a *semi-algorithm* which may go on forever or output nonsense for some input values. The concept of the GM−algorithm as a *two-state automata* was first raised in [**Sil**], we do not pursue here.

3.1. GM**−algorithm.** The GM−algorithm codifies all previous 2−generator $\mathrm{PSL}(2,\mathbf{R})$ discreteness algorithms. The overarching conceptual framework is to pair the process of *trace minimizing* with the *Poincaré/Jørgensen dichotomy* [**Gl1**]. All possible types of generating pairs are explicated into two disjoint cases: *non-intertwining* and *intertwining*, two hyperbolic generators with intersecting axes and all other cases, respectively. The Poincaré/Jørgensen dichotomy describes an *"if-else-if"* check of the necessary condition(Jørgensen) against the sufficient condition(Poincaré), with each change of generating set. The algorithm ends when a fundamental domain is built or can be determined to exist. Trace minimization works like this: starting with the ordered generating set $\{A,B\}$, replace this generating set with the generating set $\{B,A\}$ (called a *switch*), $\{A,B^{-1}\}$ (called a *inversion*), or $\{A,AB\}$ (called a *twist*) [**Gl1**]. These replacement moves are called *Nielson transformations*, they are done to minimize the absolute value of the traces of the generators or to prepare the generators for minimization. Gilman and Maskit also give a geometric interpretation of the GM−algorithm: trace minimizes corresponds to minimizing the length of the longest side of a hyperbolic triangle determined by the intersecting axes of A and B. Furthermore a *Coxeter extension* $\widetilde{\Gamma}=\langle A,B,AB\rangle$ of $\Gamma=\langle A,B\rangle$ is employed by the GM-algorithm; either both $\widetilde{\Gamma}$ and Γ are discrete or both are not discrete. Lastly, we emphasize that understanding the *commutators* of the changes in generators along with the so called *acute triangle* theorem enables the $GM-$algorithm to decide when elliptic elements will or will not have finite order. [It is a fact that hyperbolic area is independent of generating set of underlying edge identifications [**B**].] The $GM-$algorithm determines whether a fundamental domain for $\Gamma > \mathrm{PSL}(2,\mathbf{R})$ can or can not be built, in finitely many steps.

3.2. Decidability. Recall the celebrated theorem of Gödel's which may be interpreted as saying *undecidable* questions exist [**We**]. Many questions in topology and geometry we would like to build an algorithm for have been encoded by group theory into *combinatorial group theory* and *geometric group theory* (or *logic*) [**S1**]. Psychologically, a finite presentation like the one in Equation 2.1, can be thought of as describing a group like an axiomatic system [**We**]. A group is defined by generators, subject to a given set of relations. For example, the *Tietze* theorem states that two presentations define the same group if and only if there is a sequence of Nielsen moves that relate the two presentations; the *Novikov–Boone* theorem states that there are finitely presented groups with an unsolvable *word problem* [**We**]. The *word problem* asks one to give an algorithm for deciding whether a given combination of generators represents the trivial element of the group [**S1**]. It can be shown that this is an implicit property of a group, in the sense that it does not depend on the way the group is presented. We follow [**Kp1**] and take decidability to mean an algorithm *halts* either on P or on $\neg P$. Decidability questions in general depend on formulation. For example, decidability questions involving real value inputs $\mathbf{R}$ can be decided with two (not necessarily congruent) models, *BSS-computability* or *bit-computability* [**Kp1**].

3.3. Complexity. The analysis of the complexity of algorithms reveals that decidability questions are not merely dialecticism or logical piddling. The following is proved in [**Gl1**].

THEOREM 3.1. *If Γ is as described in Theorem* 1.3, *except the coefficients are restricted from* $\mathbf{R}$ *to* $\mathbf{Q}$, *then the complexity of discreteness algorithm can been seen to be linear in the number of generators.*

As a consequence of Theorem 3.1, necessary and sufficient conditions for discreteness do not exist, therefore an algorithm is the best possible solution of the discreteness problem described by Theorem 1.3 [**Gl1**]. The following is a compelling related result.

THEOREM 3.2. *[Kapovich] Discreteness is BSS-undecidable for free, rank-2 subgroups of* $\mathrm{SL}(2,\mathbf{C})$ [**Kp1**].

The proof of Theorem 3.2 involves demonstrating that boundary of the *Maskit slice* (the Maskit slice is the set of *geometrically finite* and *faithful* representations of once-punctured torus groups into $\mathrm{SL}(2,\mathbf{C})$) is NOT a countable union of algebraic sets. In particular, the boundary of the Maskit slice is not BSS-computable. This ultimately requires *the ending lamination conjecture* [**Mk**] [**BCM**]. We add one more result before concluding these brief remarks on algorithms. Using *course geometry* (in the sense of Gromov [**Gv**]) the following related discreteness problem is being prepared.

THEOREM 3.3. *[Kapovich] Given $\rho : \mathbf{F}_k \longrightarrow \mathrm{SL}(2,\mathbf{C})$ a non-elementary representation, discreteness and faithfulness is bit-computability decidable* [**Kp2**].

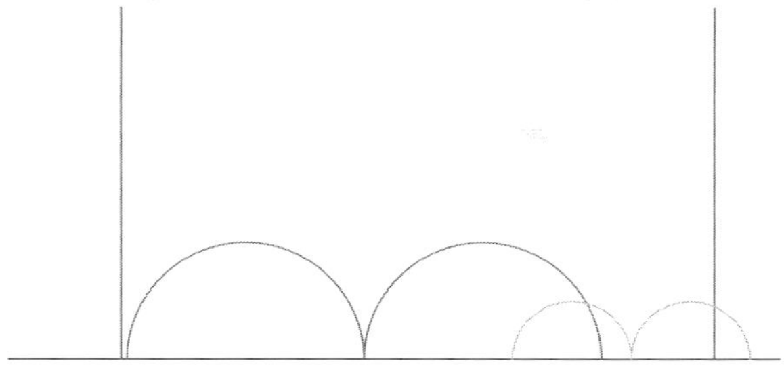

FIGURE 1. Configuration for stopping with 3 Parabolics

4. Proof of Main Theorem

The goal of this section is to give a proof of Theorem 1.4. Recall, the $4PS$-algorithm is for subgroups Γ of $\mathrm{PSL}(2,\mathbf{R})$ generated by three parabolic generators: $\langle A,B,C\rangle = \Gamma \simeq \mathbf{F}_3$, and Γ acts freely on $\mathbb{H}$. The strategy of proof for the $4PS$-algorithm is the same as that of the GM-algorithm. Namely the proof proceeds via Poincaré/Jørgensen dichotomy and trace minimizing. Either a fundamental domain for Γ can be constructed after a finite sequence of changes of the *ordered* generating set $\{A,B,C\}$, or else one is able to determine that Γ is not discrete. Figure 1 is an example of a fundamental domain of a discrete rank 3 free subgroup of $\mathrm{PSL}(2,\mathbf{R})$.

4.1. Set-up. As in the 2−generator case, conjugation of Γ by any element $A \in \mathrm{PGL}(2,\mathbf{R})$ leaves trace invariant. Hence it is possible to conjugate an element A of the generating set, by an element of $\mathrm{PGL}(2,\mathbf{R})$, say X, denoted by $A \mapsto A^X := XAX^{-1}$ so that the fixed point of A^X is ∞. Furthermore, we may conjugate A^X by suitable diagonal matrix in $\mathrm{PSL}(2,\mathbf{R})$ so that its translation length is ± 2. Next, we may conjugate the generating set by a parabolic which fixes ∞, so that the fixed point of one of the remaining generators, say B, is 0. Finally, we may conjugate the generating set by a reflection in the line from 0 to ∞ so that the fixed point of C is a positive real, label this x.

Passing to inverses if needed A,B,C can be represented by the following matrices:

$$A = \begin{bmatrix} 1 & 2 \\ 0 & 1 \end{bmatrix}, B = \begin{bmatrix} 1 & 0 \\ \frac{-2}{y} & 1 \end{bmatrix}, C = \begin{bmatrix} 1 - \frac{2x}{z} & \frac{2x^2}{z} \\ \frac{-2}{z} & 1 + \frac{2x}{z} \end{bmatrix}. \tag{4.1}$$

The significance of the triple of numbers (x,y,z) lies in the geometry of the *Ford domain*.

DEFINITION 4.1. A *Ford domain* for an element of $\mathrm{PSL}(2,\mathbf{R})$ which does not fix ∞, is a "limit" of a Dirichlet domain in which the center point z_0 goes to ∞.

In particular, a Ford domain of a general hyperbolic element $G = \begin{bmatrix} a & b \\ c & d \end{bmatrix}$ is bounded by the geodesic whose endpoints are $\frac{-d}{c} \pm \frac{1}{c}$ and the geodesic $\frac{a}{c} \pm \frac{1}{c}$. Notice that the Ford domain is symmetric with respect to the vertical line centered at $\frac{a-d}{2c}$, see Figure 2. The diameter of the compliment of one connected component of the complement of a Ford domain is $\frac{2}{|c|}$, which we will call the *Ford strength* of the transformation. [Therefore, with B and C as in Equation 4.1, y is the Ford

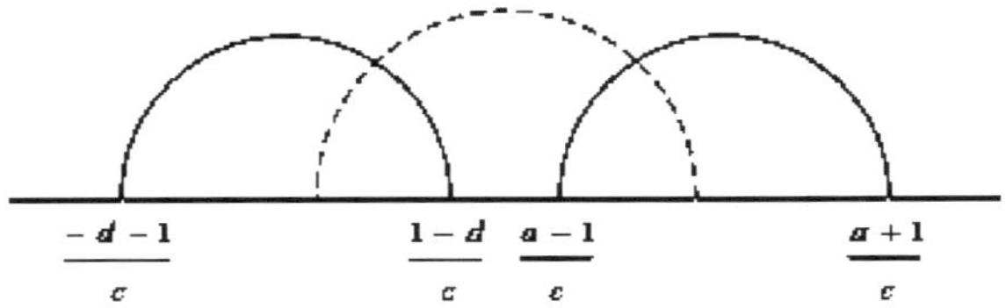

FIGURE 2. Ford domain for a generic hyperbolic element.

strength of B and z is the Ford strength of C. Also notice that the *fixed points* of the hyperbolic transformation described in Figure 2 is not the midpoints of the compliment of the Ford domain.] We will call the Euclidean distance between the connected components of the complement of the Ford domain the *inner F-distance*, given by $\frac{|\mathrm{Tr}G|-2}{|c|}$. We will call the diameter of the complement of the Ford domain the *outer F-distance*, given by $\frac{|\mathrm{Tr}G|+2}{|c|}$.

The following two Lemmas are necessary for the $4PS$-algorithm, see Step $\mathcal{D}$. Lemma 4.2 gives criteria (involving the inner and outer F-distances) for determining when A and a general non-elliptic generate a free discrete group with no elliptics. Lemma 4.3 gives a criterion (involving Ford strength) for determining if the product of some power of A and a general non-elliptic element is itself elliptic.

LEMMA 4.2. *Let* $A = \begin{bmatrix} 1 & 2 \\ 0 & 1 \end{bmatrix}$ *and* $G = \begin{bmatrix} a & b \\ c & d \end{bmatrix}$ *be any non-elliptic element of* $\mathrm{PSL}(2, \mathbf{R})$. *The following hold:*

(i) $AG, A^{-1}G, G^{-1}A$ *and* $G^{-1}A^{-1}$ *are non-elliptic* $\iff$ $|\mathrm{Tr}(G) - |2c|| \geq 2$,
(ii) $\frac{|\mathrm{Tr}(G)|+2}{|c|} < 2$, *in which case* $\langle A, G\rangle$ *generate a free discrete group with no elliptics.*

PROOF.
(i) Since G is not elliptic, $|\mathrm{Tr}G| = |a + d| \geq 2$. Without loss of generality, we can assume $a+d \geq 2$. The matrices AG, $A^{-1}G$, as well as $G^{-1}A$ and $G^{-1}A^{-1}$ have trace $(a + d) + \pm 2c$. These elements are never elliptic when $\pm 2c \geq 0$. So we only need to consider the case where the trace is $a + d - |2c|$, and $2c \neq 0$. These are elliptic if and only if $|(a + d) - |2c|| < 2$.
(ii) A Ford domain for $\langle A, G\rangle$ is given by intersection of the Ford domain of G and the half planes determined by A centered at $\frac{a-d}{2c}$. □

LEMMA 4.3. *Given* A, G *as in Lemma 4.3, if* $\frac{2}{|c|} > 1$, *then there exist some* $n \in \mathbb{N}$ *such that* A^nG *is elliptic.*

PROOF. Observe that $A^nG = \begin{bmatrix} a + 2cn & b + 2dn \\ c & d \end{bmatrix}$. Therefore,

$$\mathrm{Tr}(A^nG) = (a + d) + 2cn = \mathrm{Tr}(G) + 2cn.$$

And if $\frac{2}{|c|} > 1$ then $|2c| < 4$. Hence there exists $n \in \mathbb{N}$ such that $|\mathrm{Tr}(A^nG)| < 2$. □

4.2. Conjugation calculations. A subtlety of the $4PS-$algorithm is that Nielsen transformations do not necessarily affect trace minimization.

EXAMPLE 4.4. Let A, B, C be generating set. Consider making a combination of several Nielsen moves. For example, replace $B \mapsto ABA^{-1}$. Then the traces of the new generating set $\{A, B^A = ABA^{-1}, C\}$ all remain unchanged, as expected. However, the trace of the product B^AC *will* change and in particular it will increase.

Example 4.4 demonstrates that contrary to the case for 2-generator groups, when Nielsen moves are applied to 3-generator groups the trace of generators *may increase* after successive Nielsen moves. This motivates the following definition.

DEFINITION 4.5. For general A and B, the pair $\{A, B\}$ is *coherently oriented pairs* if and only if the $|\text{Tr}(AB)|$ is less than $|\text{Tr}(A^{-1}B)|$. Similarly, the triple $\{A, B, C\}$ is called *coherently oriented pairwise* if and only if they are coherently oriented pairs for each pair [**GK**].

Geometrically, $\{A, B, C\}$ coherently oriented pairwise implies that trace minimization affects ping-pong, as was done in the dynamical construction or Riemann surface in Example 2.8.

LEMMA 4.6. *Let P be a parabolic which does not fix ∞, and let A be in Equation 4.1. The conjugation of P by A^n does not change the Ford strength of the product P^{A^n} but translates the fixed point by $2n$.*

PROOF. This follows from a straightforward matrix multiplication. The proof is omitted. □

It is worthwhile to notice the effect of a few specific conjugations on a few special configurations of fixed points (x, y, z). The following Lemma is necessary for the $4PS$-algorithm, see Step $\mathcal{S}$.

LEMMA 4.7. *(Preparing for Minimization) Let A, B, C be as in Equation 4.1.*

(1) *If $y \neq 2x$, then the fixed point of C^B is $\frac{y^2z}{(y-2x)^2}$ and the distance between the fixed points of B and C^B is $\frac{xy}{|y-2x|}$.*
(2) *If $y = 2x$ then the fixed point of C^B is ∞. Hence either $\langle A, B, C\rangle$ is a 2−generator subgroup or $\langle A, B, C\rangle$ is not discrete.*
(3) *If $x < y - \epsilon$, for some fixed $\epsilon > 0$.*

$$z \mapsto \frac{y^2z}{(y-2(y-\epsilon))^2} = \frac{y^2z}{(-y+2\epsilon)^2} = \frac{y^2z}{(y-2\epsilon)^2} = \frac{1}{(1-\frac{2\epsilon}{y})^2}\cdot z$$

Furthermore,

$$x \mapsto \frac{yx}{y-2(y-\epsilon)|} = \frac{y}{|y-2\epsilon|}\cdot x$$

PROOF. These calculations follow from the symmetry of the configuration of three parabolic elements, like those in Equation 4.1. Conjugating B by A changes x but not y or z, hence it changes $\text{Tr}(BC)$, where ABA^{-1} is the replaces B as a generator. Also it is possible to conjugate B by A so that $x < 1$. Conjugating C by B changes x (we can assume x is positive) but some conjugation will change the order of the fixed point of B and the fixed point of C. This conjugation will also change z, replacing generators after conjugations z will be $\frac{y^2z}{(y-2x)^2}$. In particular, if $x = y$ then nothing changes. But if $x < y$ then the new z values is increased. □

One last configuration needs to be analyzed, see Step $\mathcal{D}$.

LEMMA 4.8. *Let $z \leq y \leq x \leq 1 + \epsilon$. Also assume $\frac{x^2}{|(2x-y-z)|} < 1$. Then for A, B, C as in Equation 4.1 $\langle A, B, C\rangle = \Gamma$ is discrete.*

PROOF. It suffices to construct a Ford domain for $\Gamma = \langle A, B, C\rangle$. According to the assumptions we have $p = \frac{(2x^2-xz)}{|(2x-y-z)|}$ and $p = x \cdot \frac{(2x-z)}{|(2x-y-z)|} > x$. Therefore $C(p) = \frac{(xy)}{y+z} < x$. Hence we have demonstrated a region in $\mathbb{H}$ bounded below by the geodesic whose endpoint are p and x and the geodesic whose endpoints are x and $C(p)$. Now notice that $B(C(p))$ is $\frac{(-xy)}{y+z}$. Hence the outer Ford distance of BC is less than 2 and a fundamental domain for $\langle A, B, C\rangle$ exists. □

4.3. Proof of Theorem 1.4. We begin with a summary of the conceptual strategy of the proof of Theorem 1.4. We begin by choosing a canonical starting configuration in the sense of Section 4.1. Chosen in this manner, A fixes ∞, B fixes the origin and the fixed point of C is x apart from fixed point of B. This configuration of parabolic-parabolic-parabolic generators of Γ is *canonical* up to values triples (x, y, z) which correspond to lengths between fixed points. Nielsen transformations of coherently oriented pairwise ordered generating sets always decreases trace by some amount bounded away from zero and the distance between Ford strengths decreases as the trace of the product of a coherently oriented pair decreases. Via the configurations of fixed points under Nielsen moves we can decide if we can build a Ford domain or not. The sequential procedure of the algorithm is given by the following steps:

$$\mathcal{S} \longrightarrow \mathcal{A} \longrightarrow \mathcal{D} \longrightarrow \mathcal{E}. \tag{4.2}$$

STEP $\mathcal{S}$: (Set-Up)

(1) Fix real numbers $\epsilon > 0$ and $\delta > 0$.
(2) If $x > 1 + \epsilon$, conjugate C by a power of A so that the fixed point of B and the fixed point of C are *close*; that is, within $1+\ \epsilon$. $\{A, B, C\} \mapsto \{A, B, C^{A^n}\}$. Notice that x is decreasing by at least 2ϵ, therefore the trace is seen to decrease by a fixed amount, bounded below by a function of ϵ on the order of ϵ^2.
(3) If $x < y - \delta$ conjugate C by a power of B so that the fixed point of B and the fixed point of C^B and fixed point of A are *close*. Here x is decreasing by an amount which is bounded below by order δ by Lemma 4.7. Also z is decreasing; In particular, $z \mapsto \frac{(y^2z)}{(y-2x)^2}$. So notice that if $x = y$ then nothing changes. But if $x < y$ then the new z value is decreased by an amount bounded below by order of δ.[2] Most important, $\mathrm{Tr}(AC)$ decreases by an amount bounded below by the order of δ^2.

STEP $\mathcal{A}$: (Algorithm Begins)
$\mathrm{Tr}(A) = \mathrm{Tr}(B) = \mathrm{Tr}(C) = 2$. The $(2,1)-$element of A is 0, the $(2,1)-$element of B is $\frac{-2}{y}$, and the $(2,1)-$element of C is $\frac{-2}{z}$.

(1) Calculate the matrix products; $A \cdot B, A \cdot C, B \cdot C$.
(2) $\mathrm{Tr}(BC) = -2\frac{4x^2}{yz}$, if hyperbolic $|\mathrm{Tr}(BC)| = \frac{4x^2}{yz} - 2$. *In particular*, BC is hyperbolic if and only if $x^2 > yz$.
(3) $\mathrm{Tr}(AB) = -2\frac{4}{y}$, if hyperbolic $|\mathrm{Tr}(AB)| = \frac{4}{y} - 2$.
(4) $\mathrm{Tr}(AC) = -2\frac{4}{z}$, ifhyperbolic $|\mathrm{Tr}(AC)| = \frac{4}{y} - 2$.

STEP $\mathcal{D}$: (Decisions)

By minimal positioning, the configuration is $x \geq y - \delta$ and $x \geq z - \delta$.

(1) If BC is elliptic the algorithm ends, Lemma 4.2. If BC not elliptic, then $\mathbf{x^2 \geq yz}$. Hence $x > \min(y, z)$.

(2) If $|(2x - y - z)| < yz$ then there is some n such that $A^n BC$ is elliptic by Lemma 4.3 Therefore, $|(\mathbf{2x - y - z})| \geq \mathbf{yz}$.

(3) If $z < y < x < 1$ then $1 - x \leq 1 - y$ and $1 - x \leq 1 - z$. In which case $\frac{|(x^2 - yz)|}{(2x - y - z)} < 1$ that is the inner ford strength of BC is less than 2. In which case then either ABC is elliptic or it is possible to create a fundamental domain, hence Γ is discrete, see Lemma 4.8.

STEP $\mathcal{E}$: (End Algorithm)

Now two possibilities remain observing the boundary configuration of the triple (x, y, z). Without loss of generality we assume $y > z$.

(1) If after conjugations $z < y < x < 1$ then all conditions in STEP $\mathcal{D}3$ are satisfied, hence either ABC is elliptic or it is possible to create a fundamental domain.

(2) If after conjugations $x > 1$ and $x < 1 + \epsilon$ then $2x - y - z > 0$, therefore the following holds: $\frac{(x^2 - yz)}{|(2x - y - z)|} = \frac{(x^2 - yz)}{(2x - y - z)} = \frac{(x^2 - yz)}{x - y + x - z} < \frac{(x^2 - yz)}{(x - z)} < \frac{(xy - yz)}{(x - z)} < 1$. Therefore this is Step $\mathcal{D}2$.

(3) If $(2x - y - z) < yz$ then there is some n such that $A^n BC$ is elliptic, STEP $\mathcal{D}2$.

(4) If $(2x - y - z) > yz$ then $\frac{(x^2 - yz)}{(2x - y - z)} < \frac{x^2}{xy} < 1$ and by STEP $\mathcal{D}3$, ABC is elliptic or it is possible to create a fundamental domain and Γ is discrete.

(5) If after conjugations $y > x$ and $x > y - \delta$ then conjugate C by B and either $z < y < x < 1$ and this is STEP $\mathcal{E}1$. And is decidable. Or $x > 1$ and $x < 1 + \epsilon$ and is decidable too.

(6) If after conjugations $x < z - \delta$ then conjugate B by C and either condition in STEP $\mathcal{D}2$ decidable or STEP $\mathcal{E}2$. If so then conjugate C by B and decidable.

□

5. Interpretations

We have presented the $4PS-$algorithm as a part of a sequence of related results: Theorem 2.24, Theorem 1.3, Theorem 2.31, Theorem 2.32, Theorem 3.2, Theorem 3.3. And each of these is circumscribed by the theory of (G, X)-structures on manifolds. Now we comment on why we began with this end in view.

Aside from contextualizing the $4PS-$algorithm, examples and constructions of $(G, X)-$structures give indications of the tools employed in the proof of such results. [Of course the techniques and methods of proof are often more exotic, still the fundamental ideas persist, or if not, a tincture remains in generalizations.] By way of comparing and contrasting we emphasize the following reoccurring themes: the intrinsic geometry of the hyperbolic plane or hyperbolic 3-space (e.g. the Poincaré/Jørgensen dichotomy, Coxeter extensions), utilizing informative words and dynamical techniques on character varieties (e.g. commutator words, relative character varieties), determining coordinates on character varieties (e.g. Goldman's coordinates on the $\mathrm{SL}(2, \mathbf{C})-$character varieties of rank 2 subgroups.)

Furthermore we desire to give an indication of directions of current related research. At this time, a two-generator PSL$(2, \mathbf{C})$ discreteness algorithm does not exists, but J. Gilman and L. Keen are working on this problem [**GK**]. Also in the context of decision problems, discreteness can be replaced by some other property of discrete groups, for example Γ *convex cocompact* or Γ *Anosov*, and investigate algorithmically [**Kp1**]. No complete "field guide" of $(G, X)-$structures exists.[4] By way of appealing to the mathematical precedence of advantageously reformulating problems, we emphasize that there are many correspondences for $(G, X)-$structures: $G-$representation varieties, $G-$local systems, $G-$Higgs bundles [**Gw4**] [**Gw5**] [**L1**]. These corresponding theories offer up tools and techniques which are more refined, and possibly more useful for proving theorems for discreteness algorithms. — This phenomena is illustrated quite remarkably by our discussion here of discreteness algorithms and related results for discreteness.

The quiddity of algorithms is that they are both everything and nothing. The most satisfactory solution to a discreteness problem might be an algorithm based on a computable system of coordinates for geometric structures, however searching for good coordinates might find us out to be nurturing chimera. [For example: the goal of *Jacobi Inversion problem* in classical Abel-Jacobi theory was a complete analogue of Weierstrass's $\wp$-function, the goal of the *Schottky Problem* is to compute the holomorphic invariants required to be able to distinguish holomorphic deformations of a principally polarized abelian variety.] It may not be possible to reformulate the former. An algorithm for the later exists, but the invariants are not computable. Guiding questions for creating discreteness algorithms remain: In which framework and by which means can we build *computable coordinates* for geometric structures?

QUESTION 5.1. The geometric origins of *cluster algebras* for surfaces with marked points is hyperbolic geometry and representation theory. *Cluster coordinates* of type $\mathcal{A}_n$ correspond to triangulation of an $(n-3)-$gons. Can we use *positivity* to detect discreteness, in particular discrete and faithful representations via cluster coordinates on simple hyperbolic surfaces [**AMS**]?

Acknowledgments

I am sincerely grateful to NAM and the AMS for the opportunity to be a part of this special 50th Anniversary proceedings. I would like to thank Todd Drumm and William Goldman for initiating me into this work while I was a graduate student at Howard University. I am also deeply indebted to Karen Smith and Richard Canary for mentoring me as a postdoctoral researcher at the University of Michigan. I have also benefitted greatly from affiliation with the GEAR Network, MSRI, AMS-MRC, and NAM; I happily thank my colleagues and friends for their support and encouragement. Finally, I thank each of the referees, for patiently reading earlier drafts of this paper and making many helpful comments and suggestions.

References

[A] Caleb J. Ashley, *Towards A Discreteness Algorithm For Non-Elementary Rank 3 Subgroups Of PSL(2;R)*, ProQuest LLC, Ann Arbor, MI, 2013. Thesis (Ph.D.)–Howard University. MR3187546

[4]The extraordinary book Indra's Pearls may be the prototype of such a guide. It offers stunning computer visualizations of many of the rarities in the realm of Kleinian groups [**I**].

[ABL] Caleb Ashley, Jean-Philippe Burelle, and Sean Lawton, *Rank 1 character varieties of finitely presented groups*, Geom. Dedicata **192** (2018), 1–19, DOI 10.1007/s10711-017-0281-6. MR3749420

[AMS] Ashley, C., Muller, G and Smith, K., *Cross-Ratio Coordinates for Detecting Fuchsian Groups of some Simple Hyperbolic Surfaces,* (In preparation.)

[BCM] Jeffrey F. Brock, Richard D. Canary, and Yair N. Minsky, *The classification of Kleinian surface groups, II: The ending lamination conjecture*, Ann. of Math. (2) **176** (2012), no. 1, 1–149, DOI 10.4007/annals.2012.176.1.1. MR2925381

[B] Alan F. Beardon, *The geometry of discrete groups*, Graduate Texts in Mathematics, vol. 91, Springer-Verlag, New York, 1983. MR698777

[BG] Lipman Bers and Frederick P. Gardiner, *Fricke spaces*, Adv. in Math. **62** (1986), no. 3, 249–284, DOI 10.1016/0001-8708(86)90103-9. MR866161

[B1] Bers, L., *Finite Dimensional Teichmüller Spaces and Generalizations,* Stevens and Co., New York, 1957.

[B3] Bers, L., *Uniformization, Moduli, and Kleinian Groups* Vol **5**, Number 2, AMS Bulletin, 1981.

[BBS] Lenore Blum, Mike Shub, and Steve Smale, *On a theory of computation and complexity over the real numbers: NP-completeness, recursive functions and universal machines*, Bull. Amer. Math. Soc. (N.S.) **21** (1989), no. 1, 1–46, DOI 10.1090/S0273-0979-1989-15750-9. MR974426

[CRG] R. D. Canary, D. B. A. Epstein, and P. L. Green, *Notes on notes of Thurston [MR0903850]*, Fundamentals of hyperbolic geometry: selected expositions, London Math. Soc. Lecture Note Ser., vol. 328, Cambridge Univ. Press, Cambridge, 2006, pp. 1–115. With a new foreword by Canary. MR2235710

[C2] Canary, R., *Marden's Tameness Conjecture: History and Application,* Number 2, AMS Bulletin, 1981.

[CT] Chen,L. and To, W.K. (Editors) *Geometry, Topology and Dynamics of Character Varieties,* Lecture Notes Series. IMS,NUS; World Scientific Publishing.(2012)

[D] Simon Donaldson, *Riemann surfaces*, Oxford Graduate Texts in Mathematics, vol. 22, Oxford University Press, Oxford, 2011. MR2856237

[DW] Richard Dedekind and Heinrich Weber, *Theory of algebraic functions of one variable*, History of Mathematics, vol. 39, American Mathematical Society, Providence, RI; London Mathematical Society, London, 2012. Translated from the 1882 German original and with an introduction, bibliography and index by John Stillwell. MR2962951

[E] D. B. A. Epstein, *Word processing algorithms, rewrite rules and group theory*, The mathematical revolution inspired by computing (Brighton, 1989), Inst. Math. Appl. Conf. Ser. New Ser., vol. 30, Oxford Univ. Press, New York, 1991, pp. 87–100. MR1147219

[FrKl] Felix Klein and Robert Fricke, *Lectures on the theory of elliptic modular functions. Vol. 1*, CTM. Classical Topics in Mathematics, vol. 1, Higher Education Press, Beijing, 2017. Translated from the German original [MR0247996] by Arthur M. DuPre. MR3838340

[FsKa] H. M. Farkas and I. Kra, *Riemann surfaces*, 2nd ed., Graduate Texts in Mathematics, vol. 71, Springer-Verlag, New York, 1992. MR1139765

[FM] Benson Farb and Dan Margalit, *A primer on mapping class groups*, Princeton Mathematical Series, vol. 49, Princeton University Press, Princeton, NJ, 2012. MR2850125

[Gl1] Jane Gilman, *Two-generator discrete subgroups of* $\mathrm{PSL}(2,\mathbf{R})$, Mem. Amer. Math. Soc. **117** (1995), no. 561, x+204, DOI 10.1090/memo/0561. MR1290281

[Gl2] Jane Gilman, *Algorithms, complexity and discreteness criteria in* $\mathrm{PSL}(2,\mathbf{C})$, J. Anal. Math. **73** (1997), 91–114, DOI 10.1007/BF02788139. MR1616469

[GM] J. Gilman and B. Maskit, *An algorithm for* 2*-generator Fuchsian groups*, Michigan Math. J. **38** (1991), no. 1, 13–32, DOI 10.1307/mmj/1029004258. MR1091506

[GK] Gilman, J., and Keen, Linda., *Canonical Hexagons and the* $\mathrm{PSL}(2,\mathbf{C})$ *Discreteness Problem,* 2015.

[Gw1] William M. Goldman, *Trace coordinates on Fricke spaces of some simple hyperbolic surfaces*, Handbook of Teichmüller theory. Vol. II, IRMA Lect. Math. Theor. Phys., vol. 13, Eur. Math. Soc., Zürich, 2009, pp. 611–684, DOI 10.4171/055-1/16. MR2497777

[Gw2] William M. Goldman, *Topological components of spaces of representations*, Invent. Math. **93** (1988), no. 3, 557–607, DOI 10.1007/BF01410200. MR952283

[Gw3] William M. Goldman, *Ergodic theory on moduli spaces*, Ann. of Math. (2) **146** (1997), no. 3, 475–507, DOI 10.2307/2952454. MR1491446

[Gw4] Goldman, W., *Geometric Structures on Manifolds*, `http://www.math.umd.edu/~wmg`

[Gw5] William M. Goldman, *Higgs bundles and geometric structures on surfaces*, The many facets of geometry, Oxford Univ. Press, Oxford, 2010, pp. 129–163, DOI 10.1093/acprof:oso/9780199534920.003.0008. MR2681690

[Gv] M. Gromov, *Hyperbolic groups*, Essays in group theory, Math. Sci. Res. Inst. Publ., vol. 8, Springer, New York, 1987, pp. 75–263, DOI 10.1007/978-1-4613-9586-7_3. MR919829

[I] David Mumford, Caroline Series, and David Wright, *Indra's pearls*, Cambridge University Press, Cambridge, 2015. The vision of Felix Klein; With cartoons by Larry Gonick; Paperback edition with corrections; For the 2002 edition see [MR1913879]. MR3558870

[J] Troels Jørgensen, *On discrete groups of Möbius transformations*, Amer. J. Math. **98** (1976), no. 3, 739–749, DOI 10.2307/2373814. MR427627

[K] Svetlana Katok, *Fuchsian groups*, Chicago Lectures in Mathematics, University of Chicago Press, Chicago, IL, 1992. MR1177168

[Kp1] Michael Kapovich, *Discreteness is undecidable*, Internat. J. Algebra Comput. **26** (2016), no. 3, 467–472, DOI 10.1142/S0218196716500193. MR3506344

[Kp2] Kapovich,Michael, K., *Discreteness is decidable* (In preparation)

[Ls] Roger C. Lyndon and Paul E. Schupp, *Combinatorial group theory*, Springer-Verlag, Berlin-New York, 1977. Ergebnisse der Mathematik und ihrer Grenzgebiete, Band 89. MR0577064

[L1] François Labourie, *Lectures on representations of surface groups*, Zurich Lectures in Advanced Mathematics, European Mathematical Society (EMS), Zürich, 2013. MR3155540

[MPT] Sara Maloni, Frédéric Palesi, and Ser Peow Tan, *On the character variety of the four-holed sphere*, Groups Geom. Dyn. **9** (2015), no. 3, 737–782, DOI 10.4171/GGD/326. MR3420542

[Md] Albert Marden, *Hyperbolic manifolds: An introduction in 2 and 3 dimensions*, Cambridge University Press, Cambridge, 2016. MR3586015

[M1] Bernard Maskit, *On Poincaré's theorem for fundamental polygons*, Advances in Math. **7** (1971), 219–230, DOI 10.1016/S0001-8708(71)80003-8. MR297997

[M2] Bernard Maskit, *Kleinian groups*, Grundlehren der Mathematischen Wissenschaften [Fundamental Principles of Mathematical Sciences], vol. 287, Springer-Verlag, Berlin, 1988. MR959135

[Mc] Curt McMullen, *Riemann surfaces and the geometrization of 3-manifolds*, Bull. Amer. Math. Soc. (N.S.) **27** (1992), no. 2, 207–216, DOI 10.1090/S0273-0979-1992-00313-0. MR1153266

[Mi1] John Milnor, *Hyperbolic geometry: the first 150 years*, Bull. Amer. Math. Soc. (N.S.) **6** (1982), no. 1, 9–24, DOI 10.1090/S0273-0979-1982-14958-8. MR634431

[Mi2] John W. Milnor and James D. Stasheff, *Characteristic classes*, Princeton University Press, Princeton, N. J.; University of Tokyo Press, Tokyo, 1974. Annals of Mathematics Studies, No. 76. MR0440554

[Mk] Yair Minsky, *The classification of Kleinian surface groups. I. Models and bounds*, Ann. of Math. (2) **171** (2010), no. 1, 1–107, DOI 10.4007/annals.2010.171.1. MR2630036

[Mfd] David Mumford, *Curves and their Jacobians*, The University of Michigan Press, Ann Arbor, Mich., 1975. MR0419430

[P] Henri Poincaré, *Papers on Fuchsian functions*, Springer-Verlag, New York, 1985. Translated from the French and with an introduction by John Stillwell. MR809181

[Pa] Athanase Papadopoulos, *Introduction to Teichmüller theory, old and new*, Handbook of Teichmüller theory. Vol. I–IV, IRMA Lect. Math. Theor. Phys., vol. 11, Eur. Math. Soc., Zürich, 2007, pp. 1–30, DOI 10.4171/029-1/1. MR2349667

[Rd] Reid, A., *The Arithmetic of Hyperbolic 3-Manifolds*, Springer, NY (2003).

[R] Robert Riley, *Applications of a computer implementation of Poincaré's theorem on fundamental polyhedra*, Math. Comp. **40** (1983), no. 162, 607–632, DOI 10.2307/2007537. MR689477

[Sh] Peter B. Shalen, *Representations of 3-manifold groups*, Handbook of geometric topology, North-Holland, Amsterdam, 2002, pp. 955–1044. MR1886685

[Sil] Andrew E. Silverio, *Reversing palindromic enumeration in rank-two free groups*, Exp. Math. **26** (2017), no. 3, 364–372, DOI 10.1080/10586458.2016.1180655. MR3642113

[Spr] G. Springer, *Introduction to Riemann surfaces* ISBN 0-8218-3156-9. Second Edition, 1981. AMS Chelsea.

[St] John Stallings, *Group theory and three-dimensional manifolds*, Yale University Press, New Haven, Conn.-London, 1971. A James K. Whittemore Lecture in Mathematics given at Yale University, 1969; Yale Mathematical Monographs, 4. MR0415622

[S1] John Stillwell, *The word problem and the isomorphism problem for groups*, Bull. Amer. Math. Soc. (N.S.) **6** (1982), no. 1, 33–56, DOI 10.1090/S0273-0979-1982-14963-1. MR634433

[T1] William P. Thurston, *Three-dimensional geometry and topology. Vol. 1*, Princeton Mathematical Series, vol. 35, Princeton University Press, Princeton, NJ, 1997. Edited by Silvio Levy. MR1435975

[T2] William P. Thurston, *Three-dimensional manifolds, Kleinian groups and hyperbolic geometry*, Bull. Amer. Math. Soc. (N.S.) **6** (1982), no. 3, 357–381, DOI 10.1090/S0273-0979-1982-15003-0. MR648524

[WW] Vladimir Fock, Andrey Marshakov, Florent Schaffhauser, Constantin Teleman, and Richard Wentworth, *Geometry and quantization of moduli spaces*, Advanced Courses in Mathematics. CRM Barcelona, Birkhäuser/Springer, Cham, 2016. Papers based on courses held at the Centre de Recerca Matemàtica (CRM), Bellaterra, March–June 2012; Edited by Luis Álvarez-Cónsul, Jørgen Ellegaard Andersen and Ignasi Mundet i Riera. MR3587810

[WM] Dave Witte Morris, *Introduction to arithmetic groups*, Deductive Press, [place of publication not identified], 2015. MR3307755

[Wy] Weyl, H., *The Concept of a Riemann Surface* Dover Publications, 2009.

[We] Shmuel Weinberger, *Computers, rigidity, and moduli*, M. B. Porter Lectures, Princeton University Press, Princeton, NJ, 2005. The large-scale fractal geometry of Riemannian moduli space. MR2109177

[Wt] Wolpert, S *Families of Riemann Surfaces,* CBMS Regional Conference Series in Mathematics; **113**.(2009)

Department of Mathematics, University of Michigan, Ann Arbor, Michigan 48109
Email address: cjashley@umich.edu

Contemporary Mathematics
Volume **759**, 2020
https://doi.org/10.1090/conm/759/15270

Catalan and Motzkin integral representations

Peter McCalla and Asamoah Nkwanta

Abstract. We present new proofs of eight integral representations of the Catalan numbers. Then, we create analogous integral representations of the Motzkin numbers and obtain new results. Most integral representations of counting sequences found in the literature are proved by using advanced mathematical techniques. All integral representations in this paper are proved by using standard techniques from integral calculus. Thus, we provide a simpler approach to proving integral representations of the Catalan and Motzkin numbers.

1. Introduction

The Catalan numbers [**16**] are a sequence of natural numbers that are defined as $C_n = \frac{1}{n+1}\binom{2n}{n}$ for $n \in \mathbb{Z}_{\geq 0} = \{0, 1, 2, \dots\}$. Catalan numbers appear as solutions to several combinatorics problems. See Stanley [**17, 18, 19**] for a number of combinatorial and analytical interpretations of C_n. It is closely related to the Motzkin numbers M_n [**16**] defined as [**1**]

$$M_n = \sum_{k=0}^{\lfloor n/2 \rfloor} \binom{n}{2k} C_k. \tag{1.1}$$

Finding integral representations of counting numbers is a topic of interest for its intrinsic value. The various integral representations of C_n that are given in the literature are derived from methods involving Mellin transforms, Chebyshev polynomials, the Cauchy integral formula, and other advanced techniques that are beyond elementary integral calculus. In this paper, we present several integral representations of C_n, then using two new results, prove analogous integral representations of M_n. All representations are proved by using standard techniques from integral calculus. However, the authors do not claim that this paper contains all known integral representations of C_n and M_n.

Section 2 surveys the Catalan representations and Section 3 presents the Motzkin representations. What is new in Section 2 are the proofs. Section 3 contains new results: Theorems 3.1 and 3.2 and Corollaries 3.3 and 3.5. The only known integral representations of M_n in Section 3 are Corollary 3.3(5) and (6).

2010 *Mathematics Subject Classification.* Primary 11B83, 97I50.
Key words and phrases. Catalan numbers, Motzkin numbers, integral representation, Weierstrass substitution.

They are listed in OEIS [**16**] and there are no references to proofs. The proofs of Corollary 3.3(5) and (6) are also new.

2. Integral representations of C_n

We list eight integral representations of C_n. The first representation will be proved directly and subsequent representations will be proved using equivalence. We first prove the following lemma.

LEMMA 2.1. *For $r, s \in \mathbb{Z}_{\geq 0}$ and positive real number a,*

$$\int_0^{a/2} \cos^r\left(\frac{\pi x}{a}\right)\sin^s\left(\frac{\pi x}{a}\right)dx = (-1)^r\int_{a/2}^{a} \cos^r\left(\frac{\pi x}{a}\right)\sin^s\left(\frac{\pi x}{a}\right)dx.$$

PROOF. Using the integral identity $\int_0^A f(x)\,dx = \int_0^A f(A-x)\,dx$ for continuous function f and the subtraction trigonometric identities

$$\sin(M-N) = \sin M\cos N - \cos M\sin N$$

and

$$\cos(M-N) = \cos M\cos N + \sin M\sin N$$

gives

$$\begin{aligned}\int_0^{a/2} \cos^r\left(\frac{\pi x}{a}\right)\sin^s\left(\frac{\pi x}{a}\right)dx &= \int_0^{a/2} \cos^r\left(\frac{\pi}{2}-\frac{\pi x}{a}\right)\sin^s\left(\frac{\pi}{2}-\frac{\pi x}{a}\right)dx\\ &= \int_0^{a/2} \sin^r\left(\frac{\pi x}{a}\right)\cos^s\left(\frac{\pi x}{a}\right)dx.\end{aligned}$$

Let $u = x + \frac{a}{2}$. Then by using the identities

$$\sin\left(M-\frac{\pi}{2}\right) = -\cos M \text{ and } \cos\left(M-\frac{\pi}{2}\right) = \sin M,$$

$$\begin{aligned}\int_0^{a/2} \sin^r\left(\frac{\pi x}{a}\right)\cos^s\left(\frac{\pi x}{a}\right)dx &= \int_{a/2}^{a} \sin^r\left(\frac{\pi u}{a}-\frac{\pi}{2}\right)\cos^s\left(\frac{\pi u}{a}-\frac{\pi}{2}\right)du\\ &= (-1)^r\int_{a/2}^{a} \cos^r\left(\frac{\pi u}{a}\right)\sin^s\left(\frac{\pi u}{a}\right)du. \qquad \square\end{aligned}$$

THEOREM 2.2. [**12**] *For $n \in \mathbb{Z}_{\geq 0}$,*

$$C_n = \frac{4^n}{(n+1)\pi}\int_{-1}^{1}\frac{x^{2n}}{\sqrt{1-x^2}}\,dx. \tag{2.1}$$

PROOF. We start with the identity [**6, 21**]

$$\binom{2n}{n} = \frac{2\cdot 4^n}{\pi}\int_0^{\pi/2}\cos^{2n}x\,dx.$$

Using Lemma 2.1 with $r = 2n$, $s = 0$, and $a = \pi$,

$$\binom{2n}{n} = \frac{4^n}{\pi}\int_0^{\pi}\cos^{2n}x\,dx. \tag{2.2}$$

Using the substitution $u = \cos x$, we get $dx = -\frac{1}{\sqrt{1-u^2}}\,du$ and

$$\binom{2n}{n} = \frac{4^n}{\pi}\int_{-1}^{1} \frac{u^{2n}}{\sqrt{1-u^2}}\,du.$$

The result follows by multiplying both sides by $\frac{1}{n+1}$. □

THEOREM 2.3. [**12**] *For* $n \in \mathbb{Z}_{\geq 0}$,

$$C_n = \frac{4^n}{(n+1)\pi}\int_0^1 \frac{x^n}{\sqrt{x-x^2}}\,dx. \tag{2.3}$$

PROOF. The integrand in (2.1) is an even function, so

$$\begin{aligned}
\frac{4^n}{(n+1)\pi}\int_{-1}^{1} \frac{x^{2n}}{\sqrt{1-x^2}}\,dx &= \frac{4^n}{(n+1)\pi}\int_0^1 \frac{2x^{2n}}{\sqrt{1-x^2}}\,dx \\
&= \frac{4^n}{(n+1)\pi}\int_0^1 \frac{x^{2n}}{x\sqrt{1-x^2}}\,2x\,dx \\
&= \frac{4^n}{(n+1)\pi}\int_0^1 \frac{(x^2)^n}{\sqrt{(x^2)-(x^2)^2}}\,d(x^2) \\
&= \frac{4^n}{(n+1)\pi}\int_0^1 \frac{x^n}{\sqrt{x-x^2}}\,dx.
\end{aligned}$$

□

THEOREM 2.4. [**13, 14**] *For* $n \in \mathbb{Z}_{\geq 0}$,

$$C_n = \frac{1}{2\pi}\int_0^4 x^n\sqrt{\frac{4-x}{x}}\,dx. \tag{2.4}$$

PROOF. From (2.3), let $y = 4x$. Then, we have

$$\begin{aligned}
\frac{4^n}{(n+1)\pi}\int_0^1 \frac{x^n}{\sqrt{x-x^2}}\,dx &= \frac{4^n}{(n+1)\pi}\int_0^4 \frac{\left(\frac{y}{4}\right)^n}{\sqrt{\frac{y}{4}-\frac{y^2}{16}}}\,\frac{1}{4}\,dy \\
&= \frac{1}{(n+1)\pi}\int_0^4 \frac{y^n}{\sqrt{\frac{4y-y^2}{16}}}\,\frac{1}{4}\,dy \\
&= \frac{1}{(n+1)\pi}\int_0^4 \frac{y^n}{\sqrt{4y-y^2}}\,dy \\
&= \frac{1}{(n+1)\pi}\int_0^4 \frac{y^{n+1}}{y^2\sqrt{\frac{4-y}{y}}}\,dy.
\end{aligned}$$

Using integration by parts with $u = y^{n+1}$ and $dv = \frac{1}{y^2\sqrt{\frac{4-y}{y}}}\,dy$,

$$\begin{aligned}
\frac{1}{(n+1)\pi}\int_0^4 \frac{y^{n+1}}{y^2\sqrt{\frac{4-y}{y}}}\,dy &= \frac{1}{(n+1)\pi}\left(\frac{n+1}{2}\int_0^4 y^n\sqrt{\frac{4-y}{y}}\,dy\right) \\
&= \frac{1}{2\pi}\int_0^4 y^n\sqrt{\frac{4-y}{y}}\,dy.
\end{aligned}$$

□

THEOREM 2.5. [**14**] *For $n \in \mathbb{Z}_{\geq 0}$,*

$$C_n = \frac{2^{2n+2}}{\pi} \int_0^\infty \frac{x^2}{(1+x^2)^{n+2}}\, dx. \tag{2.5}$$

PROOF. Starting from (2.4), use the reverse substitution $x = \frac{4}{1+t^2}$. Then,

$$\begin{aligned} \frac{1}{2\pi}\int_0^4 x^n \sqrt{\frac{4-x}{x}}\, dx &= \frac{1}{2\pi}\int_0^\infty \left(\frac{4}{1+t^2}\right)^n \frac{8t^2}{(1+t^2)^2}\, dt \\ &= \frac{4^n \cdot 8}{2\pi}\int_0^\infty \frac{t^2}{(1+t^2)^{n+2}}\, dt \\ &= \frac{2^{2n+2}}{\pi}\int_0^\infty \frac{t^2}{(1+t^2)^{n+2}}\, dt. \end{aligned}$$

□

THEOREM 2.6. [**22**] *For $n \in \mathbb{Z}_{\geq 0}$,*

$$C_n = \int_0^1 (2\cos(\pi x))^{2n}(2\sin^2(\pi x))\, dx. \tag{2.6}$$

PROOF. From (2.5), we use a substitution that is useful for integrating rational functions involving sine and cosine – the Weierstrass substitution $x = \tan\left(\frac{\pi t}{2}\right)$ [**20**]. By constructing a right triangle with angle $\frac{\pi t}{2}$, we get $\cos\left(\frac{\pi t}{2}\right) = \frac{1}{\sqrt{1+x^2}}$ and $\sin\left(\frac{\pi t}{2}\right) = \frac{x}{\sqrt{1+x^2}}$. Then

$$\frac{2^{2n+2}}{\pi}\int_0^\infty \frac{x^2}{(1+x^2)^{n+2}}\, dx = 2^{2n+1}\int_0^1 \cos^{2n}\left(\tfrac{\pi t}{2}\right)\sin^2\left(\tfrac{\pi t}{2}\right)\, dt.$$

Let $u = \frac{t}{2}$. Then,

$$\begin{aligned} 2^{2n+1}\int_0^1 \cos^{2n}\left(\tfrac{\pi t}{2}\right)\sin^2\left(\tfrac{\pi t}{2}\right)\, dt &= 2^{2n+2}\int_0^{1/2} \cos^{2n}(\pi u)\sin^2(\pi u)\, du \\ &= 2^{2n+1}\int_0^1 \cos^{2n}(\pi u)\sin^2(\pi u)\, du \end{aligned}$$

where the last equality is a consequence of Lemma 2.1. □

THEOREM 2.7. [**2, 11, 14**] *For $n \in \mathbb{Z}_{\geq 0}$,*

$$C_n = \frac{2^{2n+5}}{\pi}\int_0^1 \frac{x^2(1-x^2)^{2n}}{(1+x^2)^{2n+3}}\, dx. \tag{2.7}$$

PROOF. Use Lemma 2.1 to change (2.6) back to

$$2^{2n+2}\int_0^{1/2} \cos^{2n}(\pi x)\sin^2(\pi x)\, dx.$$

Let $u = \tan\left(\frac{\pi x}{2}\right)$. Then, $\cos(\pi x) = \frac{1-u^2}{1+u^2}$, $\sin(\pi x) = \frac{2u}{1+u^2}$, $dx = \frac{2}{\pi}\frac{du}{1+u^2}$, and

$$\begin{aligned} 2^{2n+2}\int_0^{1/2} \cos^{2n}(\pi x)\sin^2(\pi x)\, dx &= \frac{2^{2n+3}}{\pi}\int_0^1 \frac{(1-u^2)^{2n}}{(1+u^2)^{2n+1}}\left(\frac{2u}{1+u^2}\right)^2 du \\ &= \frac{2^{2n+5}}{\pi}\int_0^1 \frac{u^2(1-u^2)^{2n}}{(1+u^2)^{2n+3}}\, du. \end{aligned}$$

□

THEOREM 2.8. [**2, 14**] *For* $n \in \mathbb{Z}_{\geq 0}$,

$$C_n = \frac{2^{2n+1}}{\pi} \int_{-1}^{1} x^{2n} \sqrt{1-x^2}\, dx. \tag{2.8}$$

PROOF. From (2.7), use the reverse substitution $x = \sqrt{\frac{1-u}{1+u}}$ to get

$$\begin{aligned}
\frac{2^{2n+5}}{\pi} \int_0^1 \frac{x^2(1-x^2)^{2n}}{(1+x^2)^{2n+3}}\, dx &= \frac{2^{2n+5}}{\pi} \int_1^0 \frac{\frac{1-u}{1+u}\left(1-\frac{1-u}{1+u}\right)^{2n}}{\left(1+\frac{1-u}{1+u}\right)^{2n+3}} \frac{-1}{(1+u)\sqrt{1-u^2}}\, du \\
&= \frac{2^{2n+5}}{\pi} \int_0^1 \frac{\frac{1-u}{1+u}\left(\frac{2u}{1+u}\right)^{2n}}{\left(\frac{2}{1+u}\right)^{2n+3}(1+u)\sqrt{1-u^2}}\, du \\
&= \frac{2^{2n+2}}{\pi} \int_0^1 \frac{u^{2n}(1-u)(1+u)}{\sqrt{1-u^2}}\, du \\
&= \frac{2^{2n+2}}{\pi} \int_0^1 u^{2n}\sqrt{1-u^2}\, du.
\end{aligned}$$

Since the integrand is an even function, the result follows. □

REMARK 2.9. This integral representation can be proved using the beta and gamma functions and related identities [**14**].

THEOREM 2.10. [**14**] *For* $n \in \mathbb{Z}_{\geq 0}$,

$$C_n = \frac{1}{2\pi} \int_{-2}^{2} x^{2n} \sqrt{4-x^2}\, dx. \tag{2.9}$$

PROOF. From (2.8), let $u = 2x$. Then

$$\begin{aligned}
\frac{2^{2n+1}}{\pi} \int_{-1}^{1} x^{2n}\sqrt{1-x^2}\, dx &= \frac{2^{2n+1}}{\pi} \int_{-2}^{2} \left(\frac{u}{2}\right)^{2n} \sqrt{1-\left(\frac{u}{2}\right)^2}\, \frac{du}{2} \\
&= \frac{1}{\pi} \int_{-2}^{2} u^{2n} \sqrt{1-\frac{u^2}{4}}\, du \\
&= \frac{1}{2\pi} \int_{-2}^{2} u^{2n}\sqrt{4-u^2}\, du.
\end{aligned}$$

□

3. Integral representations of M_n

The next two theorems will provide a link between the representations (2.1) and (2.3)-(2.9) and Corollaries 3.3 and 3.5.

THEOREM 3.1. *Let* $n \in \mathbb{Z}_{\geq 0}$ *and* $a, b \in \mathbb{R} \cup \{\pm\infty\}$. *Suppose* C_n *can be written in the form*

$$C_n = \int_a^b \left(f(x)\right)^{2n} g(x)\, dx$$

for integrable functions f *and* g. *Then*

$$M_n = \frac{1}{2} \int_a^b \left((1+f(x))^n + (1-f(x))^n\right) g(x)\, dx. \tag{3.1}$$

PROOF. Using (1.1) and the Binomial Theorem, we have

$$\begin{aligned}M_n &= \int_a^b \sum_{k=0}^{\lfloor n/2\rfloor} \binom{n}{2k} \big(f(x)\big)^{2k} g(x)\,dx \\ &= \frac{1}{2}\int_a^b \left(\sum_{k=0}^{n}\binom{n}{k}(f(x))^k + \sum_{k=0}^{n}\binom{n}{k}(-1)^k(f(x))^k\right) g(x)\,dx \\ &= \frac{1}{2}\int_a^b ((1+f(x))^n + (1-f(x))^n)g(x)\,dx. \qquad \square\end{aligned}$$

THEOREM 3.2. *Let $n \in \mathbb{Z}_{\geq 0}$ and $a, b \in \mathbb{R} \cup \{\pm\infty\}$. Suppose C_n can be written in the form*

$$C_n = \frac{1}{n+1}\int_a^b \big(f(x)\big)^{2n} g(x)\,dx$$

for integrable functions f and g. Then

$$M_n = \int_a^b \frac{\varphi_{n+2}(x) - \varphi_{n+1}(x)}{(f(x))^2} g(x)\,dx, \tag{3.2}$$

where

$$\varphi_n(x) = \frac{1}{n}((1+f(x))^n + (1-f(x))^n - 2)$$

PROOF. We use the same techniques as the proof of Theorem 3.1; although, we need to initially convert M_n to a double integral. From (1.1), we have

$$\begin{aligned}M_n &= \int_a^b \sum_{k=0}^{\lfloor n/2\rfloor}\binom{n}{2k}\frac{1}{k+1}\big(f(x)\big)^{2k} g(x)\,dx \\ &= \int_a^b\int_0^1 \sum_{k=0}^{\lfloor n/2\rfloor}\binom{n}{2k}(f(x))^{2k}y^k g(x)\,dy\,dx\end{aligned}$$

$$\begin{aligned}&= \int_a^b\int_0^1 \sum_{k=0}^{\lfloor n/2\rfloor}\binom{n}{2k}(f(x)\sqrt{y})^{2k} g(x)\,dy\,dx \\ &= \frac{1}{2}\int_a^b g(x)\int_0^1 \left(\sum_{k=0}^{n}\binom{n}{k}(f(x)\sqrt{y})^k + \sum_{k=0}^{n}\binom{n}{k}(-1)^k(f(x)\sqrt{y})^k\right) dy\,dx \\ &= \frac{1}{2}\int_a^b g(x)\int_0^1 \left((1+f(x)\sqrt{y})^n + (1-f(x)\sqrt{y})^n\right) dy\,dx \\ &= \int_a^b g(x)\frac{1}{(f(x))^2}\left(\int_1^{1+f(x)} (u^{n+1}-u^n)\,du + \int_1^{1-f(x)} (v^{n+1}-v^n)\,dv\right) dx,\end{aligned}$$

where $u = 1 + f(x)\sqrt{y}$ and $v = 1 - f(x)\sqrt{y}$. The result follows by evaluating the inner integrals and simplifying. $\square$

COROLLARY 3.3. *For $n \in \mathbb{Z}_{\geq 0}$,*

(1) $M_n = \dfrac{1}{4\pi}\displaystyle\int_0^4 \left((1+\sqrt{x})^n + (1-\sqrt{x})^n\right)\sqrt{\frac{4-x}{x}}\,dx,$

(2) $M_n = \dfrac{2}{\pi}\displaystyle\int_0^\infty \left(\left(1+\frac{2}{\sqrt{1+x^2}}\right)^n + \left(1-\frac{2}{\sqrt{1+x^2}}\right)^n\right)\frac{x^2}{(1+x^2)^2}\,dx,$

(3) $M_n = \displaystyle\int_0^1 \left((1+2\cos\pi x)^n + (1-2\cos\pi x)^n\right)\sin^2\pi x\,dx,$

(4) $M_n = \dfrac{16}{\pi}\displaystyle\int_0^1 \frac{x^2\left((3-x^2)^n + (3x^2-1)^n\right)}{(1+x^2)^{n+3}}\,dx,$

(5) $M_n = \dfrac{2}{\pi}\displaystyle\int_{-1}^1 (1+2x)^n\sqrt{1-x^2}\,dx,$

(6) $M_n = \dfrac{1}{2\pi}\displaystyle\int_{-2}^2 (1+x)^n\sqrt{4-x^2}\,dx.$

Remark 3.4. The integral representations in Corollary 3.3(5) and (6) are presented by Peter Luschny and Paul Barry, respectively, in OEIS [**16**].

Proof. (1) Use (3.1) on (2.4) with $f(x)=\sqrt{x}$ and $g(x)=\frac{1}{2\pi}\sqrt{\frac{4-x}{x}}$.

(2) Use (3.1) on (2.5) with $f(x)=\frac{2}{\sqrt{1+x^2}}$ and $g(x)=\frac{4x^2}{\pi(1+x^2)^2}$.

(3) Use (3.1) on (2.6) with $f(x)=2\cos\pi x$ and $g(x)=2\sin^2\pi x$.

(4) Using (3.1) on (2.7) with $f(x)=\frac{2(1-x^2)}{1+x^2}$ and $g(x)=\frac{32x^2}{\pi(1+x^2)^3}$, we get

$$\begin{aligned}
M_n &= \frac{16}{\pi}\int_0^1 \left(\left(1+\frac{2-2x^2}{1+x^2}\right)^n + \left(1-\frac{2-2x^2}{1+x^2}\right)^n\right)\frac{x^2}{(1+x^2)^3}\,dx\\
&= \frac{16}{\pi}\int_0^1 \left(\left(\frac{3-x^2}{1+x^2}\right)^n + \left(\frac{3x^2-1}{1+x^2}\right)^n\right)\frac{x^2}{(1+x^2)^3}\,dx\\
&= \frac{16}{\pi}\int_0^1 \frac{x^2\left((3-x^2)^n+(3x^2-1)^n\right)}{(1+x^2)^{n+3}}\,dx.
\end{aligned}$$

(5) Using (3.1) on (2.8) with $f(x)=2x$ and $g(x)=\frac{2}{\pi}\sqrt{1-x^2}$, we get

$$\begin{aligned}
M_n &= \frac{1}{\pi}\int_{-1}^1 \left((1+2x)^n + (1-2x)^n\right)\sqrt{1-x^2}\,dx\\
&= \frac{1}{\pi}\left(\int_{-1}^1 (1+2x)^n\sqrt{1-x^2}\,dx + \int_{-1}^1 (1-2x)^n\sqrt{1-x^2}\,dx\right).
\end{aligned}$$

Perform the substitution $u=-x$ on the second integral to get the desired result.

(6) Using (3.1) on (2.9) with $f(x)=x$ and $g(x)=\frac{1}{2\pi}\sqrt{4-x^2}$, we get

$$\begin{aligned}
M_n &= \frac{1}{4\pi}\int_{-2}^2 \left((1+x)^n + (1-x)^n\right)\sqrt{4-x^2}\,dx\\
&= \frac{1}{4\pi}\left(\int_{-2}^2 (1+x)^n\sqrt{4-x^2}\,dx + \int_{-2}^2 (1-x)^n\sqrt{4-x^2}\,dx\right).
\end{aligned}$$

Perform the substitution $u=-x$ on the second integral to get the desired result.

Alternative proof of Corollary 3.3(6): From the representation in (5), perform the substitution $u = 2x$ to get

$$\begin{aligned}\frac{1}{\pi}\int_{-2}^{2}(1+u)^n\sqrt{1-\frac{u^2}{4}}\,du &= \frac{1}{\pi}\int_{-2}^{2}(1+u)^n\sqrt{\frac{4-u^2}{4}}\,du\\ &= \frac{1}{2\pi}\int_{-2}^{2}(1+u)^n\sqrt{4-u^2}\,du.\end{aligned}$$ □

COROLLARY 3.5. *For $n \in \mathbb{Z}_{\geq 0}$,*

(1) $M_n = \dfrac{1}{4\pi}\displaystyle\int_0^1 \frac{\phi_{n+2}(x)-\phi_{n+1}(x)}{x\sqrt{x-x^2}}\,dx,$

(2) $M_n = \dfrac{1}{2\pi}\displaystyle\int_{-1}^1 \frac{\psi_{n+2}(x)-\psi_{n+1}(x)}{x^2\sqrt{1-x^2}}\,dx,$

where

$$\phi_n(x) = \frac{1}{n}\big((1+2\sqrt{x})^n + (1-2\sqrt{x})^n - 2\big)$$

and

$$\psi_n(x) = \frac{1}{n}\big((1+2x)^n - 1\big).$$

PROOF. (1) Use (3.2) on (2.3) with $f(x) = 2\sqrt{x}$ and $g(x) = \frac{1}{\pi\sqrt{x-x^2}}$.

(2) Use (3.2) on (2.1) with $f(x) = 2x$ and $g(x) = \frac{1}{\pi\sqrt{1-x^2}}$. The result follows by splitting the integral, then using the substitution $u = -x$ on the integrals with the expressions $(1-2x)^{n+2}$ and $(1-2x)^{n+1}$. □

4. Conclusion

We give other known Catalan integral representations [**12**]:

$$C_n = \frac{2^{2n+1}}{(2n+1)\,\pi}\int_{-1}^{1}\frac{x^{2n+2}}{\sqrt{1-x^2}}\,dx$$

and

$$C_n = \frac{4^n}{n\pi}\int_0^1 \frac{2x^{n+1}-x^n}{\sqrt{x-x^2}}\,dx.$$

Also, from (2.2),

$$C_n = \frac{4^n}{(n+1)\pi}\int_0^{\pi}\cos^{2n}x\,dx.$$

See Qi and Guo [**14**] for integral representations of C_n and $\frac{1}{C_n}$; Gorska and Penson [**5**], for integral representations of the d-dimensional Catalan numbers; and Mlotkowski, Penson, and Zyczkowski [**7, 8, 9, 10**], for integral representations of various generalizations of standard Catalan and binomial numbers.

We have presented several integral representations of C_n and M_n. Finding integral representations of other counting numbers such as the Riordan numbers, binomial coefficients of the Pascal triangle, Fine numbers, and RNA numbers [**16**] would be of interest. See Qi, Shi, and Guo [**15**] for interesting integral representations of the little and large Schröder numbers [**16**], Dilcher [**3**] for the even Fibonacci numbers [**16**], and Glasser and Zhou [**4**] for the Fibonacci numbers [**16**].

Given the many combinatorial and analytic interpretations of C_n and M_n, an important and interesting problem worth pursuing is to find related analytical interpretations of the integral representations.

Acknowledgments

The authors would like to thank the anonymous referees, Rodney Kerby, Leon Woodson, and Guoping Zhang for useful comments on various drafts of the manuscript.

References

[1] Frank R. Bernhart, *Catalan, Motzkin, and Riordan numbers*, Discrete Math. **204** (1999), no. 1-3, 73–112, DOI 10.1016/S0012-365X(99)00054-0. MR1691863

[2] Thierry Dana-Picard, *Integral presentations of Catalan numbers and Wallis formula*, Internat. J. Math. Ed. Sci. Tech. **42** (2011), no. 1, 122–129, DOI 10.1080/0020739X.2010.519792. MR2787135

[3] Karl Dilcher, *Hypergeometric functions and Fibonacci numbers*, Fibonacci Quart. **38** (2000), no. 4, 342–363. MR1775206

[4] M. Lawrence Glasser and Yajun Zhou, *An integral representation for the Fibonacci numbers and their generalization*, Fibonacci Quart. **53** (2015), no. 4, 313–318. MR3423903

[5] Katarzyna Górska and Karol A. Penson, *Multidimensional Catalan and related numbers as Hausdorff moments*, Probab. Math. Statist. **33** (2013), no. 2, 265–274. MR3158554

[6] Thomas Koshy, *Catalan numbers with applications*, Oxford University Press, Oxford, 2009. MR2526440

[7] Wojciech Młotkowski and Karol A. Penson, *A Fuss-type family of positive definite sequences*, Colloq. Math. **151** (2018), no. 2, 289–304, DOI 10.4064/cm6894-2-2017. MR3769685

[8] Wojciech Młotkowski and Karol A. Penson, *Probability distributions with binomial moments*, Infin. Dimens. Anal. Quantum Probab. Relat. Top. **17** (2014), no. 2, 1450014, 32, DOI 10.1142/S0219025714500143. MR3212684

[9] Wojciech Młotkowski and Karol A. Penson, *The probability measure corresponding to 2-plane trees*, Probab. Math. Statist. **33** (2013), no. 2, 255–264. MR3158553

[10] Wojciech Młotkowski, Karol A. Penson, and Karol Życzkowski, *Densities of the Raney distributions*, Doc. Math. **18** (2013), 1573–1596. MR3158243

[11] Asamoah Nkwanta and Akalu Tefera, *Curious relations and identities involving the Catalan generating function and numbers*, J. Integer Seq. **16** (2013), no. 9, Article 13.9.5, 15. MR3137934

[12] Asamoah Nkwanta and Earl R. Barnes, *Two Catalan-type Riordan arrays and their connections to the Chebyshev polynomials of the first kind*, J. Integer Seq. **15** (2012), no. 3, Article 12.3.3, 19. MR2908734

[13] K. A. Penson and J.-M. Sixdeniers, *Integral representations of Catalan and related numbers*, J. Integer Seq. **4** (2001), no. 2, Article 01.2.5, 6. MR1892306

[14] F. Qi and B. Guo, *Integral representations of the Catalan numbers and their applications*, Multidisciplinary Digital Publishing Institute **5** (2017).

[15] Feng Qi, Xiao-Ting Shi, and Bai-Ni Guo, *Integral representations of the large and little Schröder numbers*, Indian J. Pure Appl. Math. **49** (2018), no. 1, 23–38, DOI 10.1007/s13226-018-0258-7. MR3777492

[16] N. J. A. Sloane, *The on-line encyclopedia of integer sequences*, Ann. Math. Inform. **41** (2013), 219–234. MR3072304

[17] Richard P. Stanley, *Catalan numbers*, Cambridge University Press, New York, 2015. MR3467982

[18] Richard P. Stanley, *Enumerative combinatorics. Vol. 2*, Cambridge Studies in Advanced Mathematics, vol. 62, Cambridge University Press, Cambridge, 1999. With a foreword by Gian-Carlo Rota and appendix 1 by Sergey Fomin. MR1676282

[19] Richard P. Stanley, *Enumerative combinatorics. Vol. 2*, Cambridge Studies in Advanced Mathematics, vol. 62, Cambridge University Press, Cambridge, 1999. With a foreword by Gian-Carlo Rota and appendix 1 by Sergey Fomin. MR1676282

[20] J. Stewart, *Calculus: Early Transcendentals, 8th ed.*, Cengage Learning, 2015.

[21] J. Wiener, *Integrals of* $\cos^{2n} x$ *and* $\sin^{2n} x$, College Math. J. **31** (2000), 60–61.
[22] Q. Yuan, *The Catalan numbers, regular languages, and orthogonal polynomials*, posted online at `http://qchu.wordpress.com/2009/06/07/`, 2009.

Department of Mathematics, Morgan State University, Baltimore, Maryland 21251
Email address: `peter.mccalla@morgan.edu`

Department of Mathematics, Morgan State University, Baltimore, Maryland 21251
Email address: `asamoah.nkwanta@morgan.edu`

Contemporary Mathematics
Volume **759**, 2020
https://doi.org/10.1090/conm/759/15271

The research of seven students at Howard University

Neil Hindman

ABSTRACT. Results from the research of my final seven Ph.D. students at Howard University are discussed.

1. Introduction

During my 37 years on the faculty of the Department of Mathematics at Howard University, I have been fortunate to be asked to serve as dissertation advisor by twenty students. From the point of view of the National Association of Mathematicians, it is probably worth noting that ten of these have been black females and eight have been black males. (I do not write "African American" because not all of them were Americans.) On the occasion of my obituary conference in 2008, when I turned 65 years old, I wrote a survey of the dissertation research of all of my students up to that time [**4**]. This paper consists of a survey of the remaining seven students. (Since I have now retired and do not currently have any students, my list of students is presumably complete.)

I will not attempt to give a summary of each of the dissertations being discussed. Rather, I will attempt to pick out results that give a flavor of the major thrust of the dissertation. This will usually mean that the strongest results of the dissertation do not even get mentioned, since these strong results tend to be quite complicated.

All of my students' dissertations have involved results in Ramsey Theory or the algebraic structure of the Stone-Čech compactification, βS, of a discrete semigroup S, or both. Most of the dissertations involving both Ramsey Theory and algebra involve applications of algebraic results to obtain results in Ramsey Theory. Two of the dissertations being discussed here have algebraic results and Ramsey Theoretic results that are essentially unrelated.

This survey is organized by subject matter. Section 2 will deal with purely combinatorial Ramsey Theory. Section 3 will involve Ramsey Theoretic results related to the algebraic structure of βS. And Section 4 will consist of purely algebraic results about S and βS. (The reason for the ordering of Sections 3 and 4 is that one of the results in Section 4 is motivated by a result in Section 3.)

2. Ramsey Theory

Ramsey Theory gets its name from the following theorem. We let $\mathbb{N}$ be the set of positive integers. Given a set X and $k \in \mathbb{N}$, $[X]^k = \{A \subseteq X : |A| = k\}$.

2010 *Mathematics Subject Classification.* Primary 05D10, 22A15, 54D80.

THEOREM 2.1 (Ramsey's Theorem). *Let X be a set, let $k, r \in \mathbb{N}$, and assume that $[X]^k = \bigcup_{i=1}^r C_i$. There exist $i \in \{1, 2, \ldots, r\}$ and an infinite set $Y \subseteq X$ such that $[Y]^k \subseteq C_i$.*

PROOF. [**15**, Theorem A]. □

In the alternate chromatic terminology, if $[X]^k$ is r-colored, there is an infinite set $Y \subseteq X$ such that $[Y]^k$ is *monochromatic*.

One of the major results in Ramsey Theory is the Hales-Jewett Theorem. This involves the notion of a *free semigroup*. We shall stick to an informal treatment of free semigroups. For a more formal treatment see [**6**, Definition 1.3].

DEFINITION 2.2. Let A be a nonempty set. The *free semigroup S over the alphabet A* is the set of all finite sequences in A with the operation of concatenation. The members of S are called *words*.

For example, if $A = \{1, 2, 3, 4\}$ and S is the free semigroup over A, then $x = 12213$ and $y = 21423$ are members of S and $xy = 1221321423$.

DEFINITION 2.3. Let A be a nonempty set.

(a) A *variable word* over A is a word over $A \cup \{v\}$ in which v occurs, where v is a *variable* which is not a member of A.

(b) If w is a variable word over A and $a \in A$, then $w(a)$ is the result of replacing each occurrence of v in w by a.

For example, if $A = \{1, 2, 3, 4\}$ and $w = 13v2v3$, then $w(1) = 131213$ and $w(4) = 134243$.

THEOREM 2.4 (Hales-Jewett Theorem). *Let A be a finite nonempty set, let S be the free semigroup over A, let $r \in \mathbb{N}$, and let $S = \bigcup_{i=1}^r C_i$. There exist $i \in \{1, 2, \ldots, r\}$ and a variable word w over A such that $\{w(a) : a \in A\} \subseteq C_i$.*

PROOF. [**3**, Theorem 1]. □

The set $\{w(a) : a \in A\}$ is frequently referred to as a *combinatorial line*.

Another, by now reasonably old, result in Ramsey Theory is the Finite Products Theorem. For a set X, we let $\mathcal{P}_f(X)$ be the set of finite nonempty subsets of X. Given a semigroup $(S, \cdot)$ and a sequence $\langle x_n \rangle_{n=1}^\infty$ in S, we let $FP(\langle x_n \rangle_{n=1}^\infty) = \{\prod_{t \in F} x_t : F \in \mathcal{P}_f(\mathbb{N})\}$ where $\prod_{t \in F} x_t$ is computed in increasing order of indices. If the operation in S is denoted by $+$, we write $FS(\langle x_n \rangle_{n=1}^\infty) = \{\sum_{t \in F} x_t : F \in \mathcal{P}_f(\mathbb{N})\}$.

THEOREM 2.5 (Finite Products Theorem). *Let $(S, \cdot)$ be a semigroup, let $r \in \mathbb{N}$, and let $S = \bigcup_{i=1}^r C_i$. There exist $i \in \{1, 2, \ldots, r\}$ and a sequence $\langle x_n \rangle_{n=1}^\infty$ in S such that $FP(\langle x_n \rangle_{n=1}^\infty) \subseteq C_i$.*

PROOF. [**6**, Corollary 5.9]. □

The last Ramsey Theoretic topic that is addressed by the students' dissertations is *image partition regularity* of matrices.

DEFINITION 2.6. Let $u, v \in \mathbb{N}$ and let A be a $u \times v$ matrix with integer entries.

(a) The matrix A is *image partition regular over* $\mathbb{N}$ if and only if, whenever $r \in \mathbb{N}$ and $\mathbb{N} = \bigcup_{i=1}^r C_i$, there exist $i \in \{1, 2, \ldots, r\}$ and $\vec{x} \in \mathbb{N}^v$ such that $A\vec{x} \in (C_i)^u$.

(b) The matrix A is *weakly image partition regular over* $\mathbb{N}$ if and only if, whenever $r \in \mathbb{N}$ and $\mathbb{N} = \bigcup_{i=1}^{r} C_i$, there exist $i \in \{1, 2, \ldots, r\}$ and $\vec{x} \in \mathbb{Z}^v$ such that $A\vec{x} \in (C_i)^u$.

Numerous characterizations of image partition regular matrices are known. See [**6**, Theorem 15.24]. To illustrate, the fact that the matrix

$$\begin{pmatrix} 1 & 0 \\ 1 & 1 \\ 1 & 2 \\ 1 & 3 \\ 1 & 4 \end{pmatrix}$$

is image partition regular says that whenever $\mathbb{N}$ is divided into finitely many classes, one of those classes contains a length 5 arithmetic progression. This is a special case of van der Waerden's Theorem [**17**].

Henry Jordan. Let $X = \{1, 2, 3\}^4$. We will denote the members of X without commas or parentheses. So, for example, we will write 1213 instead of $(1, 2, 1, 3)$. It was shown in [**7**] that if X is 2-colored, there is a monochromatic line. That is, there is a variable word w over $\{1, 2, 3\}$ such that $\{w(1), w(2), w(3)\}$ is monochromatic.

DEFINITION 2.7. A set $Y \subseteq X$ is a *Hales-Jewett set* if and only whenever Y is 2-colored, there must be a monochromatic combinatorial line.

Note that X has 81 members. In his dissertation [**8**], Dr. Jordan produced by analytic methods a set with 69 members which is not a Hales-Jewett set. Specifically, if

$$Y = X \setminus \{1233, 1323, 1332, 2133, 2313, 2331, 3123, 3132, 3213, 3231, 3312, 3321\}$$

then there is a 2-coloring of Y with no monochromatic lines.

He then used computer techniques (programming in Pascal) to produce a minimal Hales-Jewett set with 44 elements. For each possible candidate, the computer searched for a 2-coloring with no monochromatic line. If none was found, then the set was a Hales-Jewett set. Since there are 2^{44} 2-colorings of a 44 element set, the program obviously can't just try every one. But, for example, if 1111 and 1112 have been assigned to the same color and one is trying to avoid a monochromatic line, then 1113 must be assigned to the other color. Using such a program repeatedly, he found that if

$$\begin{aligned} Y = \; & \{1111, 2222, 3333, 1222, 1112, 1121, 1122, 2111, 1211, 1333, 1113, 1313, 1133, \\ & 1131, 1223, 1233, 1323, 2223, 2212, 2232, 2122, 3332, 3313, 3323, 3133, 3233, \\ & 2112, 2121, 2323, 2211, 2233, 3113, 3223, 3131, 3322, 2131, 2133, 3112, 3122, \\ & 1123, 2213, 2333, 3111, 3222\} \end{aligned}$$

then (1) whenever Y is 2-colored, there must be a monochromatic combinatorial line and (2) whenever any one of the 44 elements of Y is deleted, there is a 2-coloring without a monochromatic combinatorial line.

Dev Phulara.

DEFINITION 2.8. Let $a, r \in \mathbb{N}$. Then $SP_2(a, r)$ is the first $n \in \mathbb{N}$, if such exists, such that whenever $\{a, a+1, \ldots, n\}$ is r-colored, there exist x and y with $a \leq x < y$ such that $\{x + y, xy\}$ is monochromatic. If no such n exists, then $SP_2(a, r) = \infty$.

It is an old result of Ron Graham, never published by him, that $SP_2(a,2)$ is finite for all a, and his argument allows one to compute upper bounds.

In the combinatorial portion of his dissertation [**13**], Dr. Phulara established that for all $a \in \mathbb{N}$, $SP_2(a,n) \geq a^2(a+\lfloor 2\sqrt{a}\rfloor)$ and computed some upper bounds on $SP_2(a,2)$ that were slight improvements over the ones computed using Graham's argument.

Using a sophisticated computer program written in Pascal, Dr. Phulara also computed the exact value of $SP_2(a,2)$ for every $a \in \{1,2,\ldots,105\}$. For example, $SP_2(105,2) = 1543500$. To establish this, his program needed to find a 2-coloring of $\{105,106,\ldots,1543499\}$ without any x and y with $105 \leq x < y$ such that $x+y$ and xy were the same color. And it had to establish that for any 2-coloring of $\{105,106,\ldots,1543500\}$ some such x and y must exist.

One fascinating fact arose from the computation of these exact values. This was that in every computed case, $SP_2(a,2)$ is divisible by a^2. Nobody has been able to prove that $SP_2(a,2)$ is always divisible by a.

Kendra Pleasant. The notion of weakly image partition regular is weaker than the notion of image partition regular. For example, it was shown in [**5**] that the matrix

$$\begin{pmatrix} 1 & -1 \\ 3 & 2 \\ 4 & 6 \end{pmatrix}$$

is weakly partition regular but not partition regular.

In the combinatorial portion of her dissertation [**14**], Dr. Pleasant proved the following theorem. This shows that the relevant images obtained for a weakly image partition regular matrix can also be obtained for an image partition regular matrix.

THEOREM 2.9. *Let $u, v, n \in \mathbb{N}$ and let A be a $u \times v$ matrix of rank n with integer entries. There is a $u \times n$ matrix B with integer entries such that*

$$\{A\vec{k} : \vec{k} \in \mathbb{Z}^v\} \cap \mathbb{N}^u = \{B\vec{x} : \vec{x} \in \mathbb{N}^n\} \cap \mathbb{N}^u\,.$$

In particular, if A is weakly image partition regular over $\mathbb{N}$, then B is image partition regular over $\mathbb{N}$.

PROOF. [**14**, Theorem 22]. □

3. Ramsey Theory and βS

We take the Stone-Čech compactification βX of a discrete space X to be the set of ultrafilters on X, with the principal ultrafilters being identified with the points of X. Given a discrete semigroup $(S,\cdot)$, the operation extends to βS so that $(\beta S,\cdot)$ is a right topological semigroup, meaning that the function $q \mapsto q\cdot p$ from βS to itself is continuous for each $p \in \beta S$. Further, the function $q \mapsto x \cdot q$ is continuous for each $x \in S$.

Any compact Hausdorff right topological semigroup T has a smallest two sided ideal $K(T)$ and $K(T)$ has idempotents. An idempotent in $K(T)$ is said to be a *minimal* idempotent. There is an important relationship between idempotents and sets of finite products (or finite sums).

THEOREM 3.1. *Let $(S,\cdot)$ be a semigroup.*

(1) *If p is an idempotent in $(\beta S,\cdot)$ and $A \in p$, then there is a sequence $\langle x_n\rangle_{n=1}^\infty$ such that $FP(\langle x_n\rangle_{n=1}^\infty) \subseteq A$.*

(2) *If* $\langle x_n\rangle_{n=1}^\infty$ *is a sequence in* S, *then* $\bigcap_{m=1}^\infty c\ell_{\beta S}\big(FP(\langle x_n\rangle_{n=m}^\infty)\big)$ *is a subsemigroup of* $(\beta S,\cdot)$ *and there is an idempotent in this subsemigroup. In particular, there is an idempotent* $p\in\beta S$ *such that* $FP(\langle x_n\rangle_{n=1}^\infty)\in p$.

PROOF. (1) [**6**, Theorem 5.8]. (2) [**6**, Lemma 5.11]. □

DEFINITION 3.2. Let $(S,\cdot)$ be a semigroup and let $\langle x_n\rangle_{n=1}^\infty$ be a sequence in S. Then $\langle y_n\rangle_{n=1}^\infty$ is a *product subsystem* of $\langle x_n\rangle_{n=1}^\infty$ if and only if there exists a sequence $\langle F_n\rangle_{n=1}^\infty$ in $\mathcal{P}_f(\mathbb{N})$ such that for each n, $\max F_n<\min F_{n+1}$ and $y_n=\prod_{t\in F_n}x_t$.

The analogous notion for a semigroup written additively is called a *sum subsystem*.

A very important notion, both algebraically and combinatorially, is the notion of *central* sets.

DEFINITION 3.3. Let $(S,\cdot)$ be a semigroup and let $C\subseteq S$. Then C is *central* in S if and only if there is an idempotent $p\in K(\beta S,\cdot)$ such that $C\in p$.

The original Central Sets Theorem, proved using a different but equivalent definition of central, is the following.

THEOREM 3.4 (Original Central Sets Theorem). *Let* C *be a central subset of* $(\mathbb{N},+)$, *let* $k\in\mathbb{N}$, *and for each* $i\in\{1,2,\ldots,k\}$, *let* $\langle y_{i,n}\rangle_{n=1}^\infty$ *be a sequence in* $\mathbb{Z}$. *There exists sequences* $\langle a_n\rangle_{n=1}^\infty$ *in* $\mathbb{N}$ *and* $\langle H_n\rangle_{n=1}^\infty$ *in* $\mathcal{P}_f(\mathbb{N})$ *such that*

(1) *for each* $n\in\mathbb{N}$, $\max H_n<\min H_{n+1}$ *and*
(2) *for each* $F\in\mathcal{P}_f(\mathbb{N})$ *and each* $i\in\{1,2,\ldots,k\}$, $\sum_{n\in F}(a_n+\sum_{t\in H_n}y_{i,t})\in C$.

PROOF. [**2**, Proposition 8.21]. □

In [**1**] the Central Sets Theorem was extended so as to handle all sequences at once in arbitrary semigroups. The extension in commutative semigroups is reasonably simple to state.

THEOREM 3.5 (Commutative Central Sets Theorem). *Let* $(S,+)$ *be a commutative semigroup, let* $\mathbb{F}$ *be the set of sequences in* S, *and let* C *be a central subset of* S. *There exist functions* $\alpha:\mathcal{P}_f(\mathbb{F})\to S$ *and* $H:\mathcal{P}_f(\mathbb{F})\to\mathcal{P}_f(\mathbb{N})$ *such that*

(1) *if* $F,G\in\mathcal{P}_f(\mathbb{F})$ *and* $F\subsetneq G$, *then* $\max H(F)<\min H(G)$ *and*
(2) *whenever* $m\in\mathbb{N}$, $G_1,G_2,\ldots,G_m\in\mathcal{P}_f(\mathbb{F})$ *such that* $G_1\subsetneq G_2\subsetneq\ldots\subsetneq G_m$, *and for each* $i\in\{1,2,\ldots,m\}$, $f_i\in G_i$, *one has* $\sum_{i=1}^m\big(\alpha(G_i)+\sum_{t\in H(G_i)}f_i(t)\big)\in C$.

PROOF. [**1**, Theorem 2.2]. □

The Central Sets Theorem for arbitrary semigroups is more complicated because the elements a_n must be split into several parts.

DEFINITION 3.6. Let $m\in\mathbb{N}$. Then $\mathcal{I}_m=\{(H_1,H_2,\ldots,H_m):$ each $H_i\in\mathcal{P}_f(\mathbb{N})$ and if $i\in\{1,2,\ldots,m-1\}$, then $\max H_i<\min H_{i+1}\}$.

THEOREM 3.7 (General Central Sets Theorem). *Let* $(S,\cdot)$ *be a semigroup, let* $\mathbb{F}$ *be the set of sequences in* S, *and let* C *be a central subset of* S. *There exist functions* $m:\mathcal{P}_f(\mathbb{F})\to\mathbb{N}$, $\alpha\in\times_{F\in\mathcal{P}_f(\mathbb{F})}S^{m(F)+1}$, *and* $H\in\times_{F\in\mathcal{P}_f(\mathbb{F})}\mathcal{I}_{m(F)}$ *such that*

(1) *if* $F, G \in \mathcal{P}_f(\mathbb{F})$ *and* $F \subsetneq G$, *then* $\max H(F)\big(m(F)\big) < \min H(G)(1)$ *and*
(2) *whenever* $n \in \mathbb{N}$, $G_1, G_2, \ldots, G_n \in \mathcal{P}_f(\mathbb{F})$ *such that* $G_1 \subsetneq G_2 \subsetneq \ldots \subsetneq G_n$, *and for each* $i \in \{1, 2, \ldots, n\}$, $f_i \in G_i$, *one has*
$\prod_{i=1}^{n} \big(\prod_{j=1}^{m(G_i)} \big(\alpha(G_i)(j) \cdot \prod_{t \in H(G_i)(j)} f_i(t)\big)\big)\alpha(G_i)(m(G_i)+1) \in C$.

PROOF. [**1**, Corollary 3.10]. □

Kendall Williams. The main results of Dr. Williams' dissertation [**18**] are quite complicated to state. I will present here a simple special case which should convey the flavor of some of these results.

DEFINITION 3.8. Let $k \in \mathbb{N}$, let $\langle a_i \rangle_{i=1}^{k}$ be a sequence in $\mathbb{N}$ such that, if $i < k$, $a_i \neq a_{i+1}$, and let $\langle x_n \rangle_{n=1}^{\infty}$ be a sequence in $\mathbb{N}$. Then $MT(\vec{a}, \langle x_n \rangle_{n=1}^{\infty}) = \{\sum_{i=1}^{k} a_i \sum_{t \in F_i} x_t : F_1, F_2, \ldots, F_k \in \mathcal{P}_f(\mathbb{N}) \text{ and } \max F_i < \min F_{i+1} \text{ if } i < k\}$.

The set $MT(\vec{a}, \langle x_n \rangle_{n=1}^{\infty})$ is the *Milliken-Taylor system* determined by $\vec{a}$ and $\langle x_n \rangle_{n=1}^{\infty}$. The Milliken-Taylor systems are so named because of the relationship with the Milliken-Taylor Theorem ([**10**, Theorem 2.2] and [**16**, Lemma 2.2]).

THEOREM 3.9. *Let* $k \in \mathbb{N}$, *let* $\langle a_i \rangle_{i=1}^{k}$ *be a sequence in* $\mathbb{N}$ *such that, if* $i < k$, $a_i \neq a_{i+1}$, *and let* $\langle x_n \rangle_{n=1}^{\infty}$ *be a sequence in* $\mathbb{N}$. *Let* p *be an idempotent in* $\bigcap_{m=1}^{\infty} c\ell_{\beta\mathbb{N}}\big(FS(\langle x_n \rangle_{n=m}^{\infty})\big)$, *and let* $A \in a_1 p + a_2 p + \ldots a_k p$. *There exists a sum subsystem* $\langle y_n \rangle_{n=1}^{\infty}$ *of* $\langle x_n \rangle_{n=1}^{\infty}$ *such that* $MT(\vec{a}, \langle y_n \rangle_{n=1}^{\infty}) \subseteq A$.

PROOF. [**6**, Theorem 17.31]. □

The following is a special case of one of Dr. Williams' results. If one stopped at the second term, it would be easy to prove. As is, it is not at all easy.

THEOREM 3.10. *Let* p, q *be idempotents in* $(\beta\mathbb{N}, +)$ *and* $\langle x_n \rangle_{n=1}^{\infty}$ *and* $\langle y_n \rangle_{n=1}^{\infty}$ *be sequences in* $\mathbb{N}$ *such that*

$$p \in \bigcap_{m=1}^{\infty} c\ell_{\beta\mathbb{N}}\big(FS(\langle x_n \rangle_{n=m}^{\infty})\big) \text{ and } q \in \bigcap_{m=1}^{\infty} c\ell_{\beta\mathbb{N}}\big(FS(\langle y_n \rangle_{n=m}^{\infty})\big).$$

Let $a_1, a_2, a_3 \in N$ *and* $A \in a_1 p + a_2 q + a_3 p$. *Then there exist sum subsystems* $\langle u_n \rangle_{n=1}^{\infty}$ *of* $\langle x_n \rangle_{n=1}^{\infty}$ *and* $\langle v_n \rangle_{n=1}^{\infty}$ *of* $\langle y_n \rangle_{n=1}^{\infty}$ *such that* $\{a_1 \sum_{t \in F_1} u_t + a_2 \sum_{t \in F_2} v_t + a_3 \sum_{t \in F_3} u_t : F_1, F_2, F_3 \in \mathcal{P}_f(\mathbb{N}), \max F_1 < \min F_2, \text{ and } \max F_2 < \min F_3\} \subseteq A$.

PROOF. [**18**, Theorem 3.4]. □

The difficulty of this theorem is in the choice of the subsystem $\langle u_n \rangle_{n=1}^{\infty}$ of $\langle x_n \rangle_{n=1}^{\infty}$. To get a small idea of the complexity of Dr. Williams main results, he obtains similar conclusions for expressions like $-\frac{2}{3}p_1 p_3 + p_3 p_2 + 3p_1 p_2 p_3 + p_2 p_1$ where the multiplication is in $(\beta\mathbb{N}, \cdot)$ and the addition is in $(\beta\mathbb{N}, +)$.

John H. Johnson. Dr. Johnson has several results in his dissertation [**9**], but to my mind the most impressive is his immense simplification of the general Central Sets Theorem, Theorem 3.7.

DEFINITION 3.11. Let $m \in \mathbb{N}$. Then $\mathcal{J}_m = \{(t_1, t_2, \ldots, t_m) \in \mathbb{N}^m : t_1 < t_2 < \ldots < t_m\}$.

Dr. Johnson's simplification replaces the finite sets H_i by single elements.

THEOREM 3.12 (Simplified General Central Sets Theorem). *Let* $(S,\cdot)$ *be a semigroup, let* $\mathbb{F}$ *be the set of sequences in* S*, and let* C *be a central subset of* S*. There exist functions* $m:\mathcal{P}_f(\mathbb{F})\to\mathbb{N}$, $\alpha\in\times_{F\in\mathcal{P}_f(\mathbb{F})}S^{m(F)+1}$, *and* $\tau\in\times_{F\in\mathcal{P}_f(\mathbb{F})}\mathcal{J}_{m(F)}$ *such that*

(1) *if* $F,G\in\mathcal{P}_f(\mathbb{F})$ *and* $F\subsetneq G$, *then* $\tau(F)\big(m(F)\big)<\tau(G)(1)$ *and*
(2) *whenever* $n\in\mathbb{N}$, $G_1,G_2,\ldots,G_n\in\mathcal{P}_f(\mathbb{F})$ *such that* $G_1\subsetneq G_2\subsetneq\ldots\subsetneq G_n$ *and for each* $i\in\{1,2,\ldots,n\}$, $f_i\in G_i$, *one has* $\prod_{i=1}^{n}\big(\prod_{j=1}^{m(G_i)}\big(\alpha(G_i)(j)\cdot f_i\big(\tau(G_i)(j)\big)\big)\big)\alpha(G_i)(m(G_i)+1)\in C$.

If you think this is still pretty complicated, you are correct, but Theorem 3.12 is less complicated than Theorem 3.7. Notice that there is one fewer product in the computation. The product $\prod_{t\in H(G_i)(j)}f_i(t)$ is replaced by the single term $f_i\big(\tau(G_i)(j)\big)$.

Dev Phulara. In the algebraic portion of his dissertation, Dr. Phulara extended both the commutative Central Sets Theorem and the general Central Sets Theorem. Both of these results deal with arbitrary central sets, so they could be phrased as "let p be a minimal idempotent in βS and let $C\in p$." Dr. Phulara's extensions begin "let p be a minimal idempotent in βS and let $\langle C_n\rangle_{n=1}^{\infty}$ be a sequence of members of p." The rest of the statements remain the same except that the products which were guaranteed to be in C, are now guaranteed to be in C_k, where $k=|G_1|$.

Kendra Pleasant. In the algebraic portion of her dissertation, Dr. Pleasant extended the commutative Central Sets Theorem to *partial semigroups*. A *partial semigroup* is a pair $(S,*)$, where $*$ is a binary operation defined on some, but not necessarily all, elements of $S\times S$ with the property that for all $x,y,z\in S$, $(x*y)*z=x*(y*z)$ in the sense that if one side is defined, so is the other, and they are equal. The partial semigroup $(S,*)$ is *adequate* provided that for any $F\in\mathcal{P}_f(S)$, there is some $y\in S$ such that $x*y$ is defined for all $x\in F$. If $(S,*)$ is adequate, then $\delta S=\bigcap_{F\in\mathcal{P}_f(S)}c\ell_{\beta S}\{y\in S:(\forall x\in F)(x*y\text{ is defined})\}$ is a semigroup. As a compact Hausdorff right topological semigroup, δS has a smallest ideal so the definition of central sets as members of minimal idempotents makes sense.

Dr. Pleasant established that the commutative Central Sets Theorem remains valid in an adequate partial semigroup almost verbatim. The distinction is that instead of taking $\mathbb{F}$ as the set of all sequences in S, she defines $\mathbb{F}$ to be the set of all *adequate* sequences in S. A sequence $\langle f(n)\rangle_{n=1}^{\infty}$ in a partial semigroup $(S,+)$ is *adequate* if and only if (1) for each $F\in\mathcal{P}_f(\mathbb{N})$, $\sum_{t\in F}f(t)$ is defined, and (2) for each $L\in\mathcal{P}_f(S)$, there exists $m\in\mathbb{N}$ such that $x+y$ is defined for all $x\in L$ and all $y\in FS(\langle x_n\rangle_{n=m}^{\infty})$. With this change in the definition of $\mathbb{F}$, the commutative Central Sets Theorem for adequate partial semigroups is verbatim the same as Theorem 3.5.

4. Algebra of βS

Two of the dissertations being surveyed deal with purely algebraic questions about S and βS, though the notion of J-sets is motivated by the Central Sets Theorem.

One of the major unsolved problems about the algebra of $(\beta\mathbb{N},+)$ is whether there is a nontrivial continuous homomorphism from $\beta\mathbb{N}$ to $\mathbb{N}^* = \beta\mathbb{N}\setminus\mathbb{N}$. (It is known that any such continuous homomorphism must have finite range.)

DEFINITION 4.1. Let $(S,\cdot)$ be a semigroup and let $\mathbb{F}$ be the set of sequences in S.

(a) Let $A\subseteq S$. Then A is a J-set in S if and only if for each $F\in\mathcal{P}_f(\mathbb{F})$, there exist $m\in\mathbb{N}$, $a\in S^{m+1}$, and $t\in\mathcal{J}_m$ such that for each $f\in F$, $a(1)\cdot f\big(t(1)\big)\cdot a(2)\cdot f\big(t(2)\big)\cdots a(m)\cdot f\big(t(m)\big)\cdot a(m+1)\in A$.

(b) $J(S)=\{p\in\beta S:(\forall A\in p)(A \text{ is a } J\text{-set})\}$.

It is known [**6**, Theorem 14.15.1] that A satisfies the conclusion of the general Central Sets Theorem (Theorem 3.12) if and only if there is an idempotent in $J(S)\cap c\ell_{\beta S}A$.

Kourtney Fulton Miller. In her dissertation [**11**], Dr. Miller dealt with the free semigroup S_ω on the distinct generators $\{a_n : n\in\mathbb{N}\}$ and for $n\in\mathbb{N}$ the free semigroup S_n on the generators $\{a_1,a_2,\ldots,a_n\}$. She also utilized the ordering of idempotents. Given a semigroup $(S,\cdot)$ and idempotents $p,q\in\beta S$, $p\leq q$ if and only if $p=p\cdot q=q\cdot p$ and $p<q$ if and only if $p\leq q$ and $p\neq q$. An idempotent p is minimal with respect to this ordering if and only if $p\in K(\beta S)$.

The following simple result allowed her to effectively address the problem of existence of continuous homomorphisms from βS_ω to $S_\omega^*=\beta S_\omega\setminus S_\omega$.

THEOREM 4.2. *There exists a sequence $\langle q_n\rangle_{n=1}^\infty$ of idempotents in S_ω^* such that for each $n\in\mathbb{N}$, $q_n\in K(\beta S_n)$ and $q_{n+1}<q_n$.*

PROOF. [**11**, Lemma 2.9]. □

Dr. Miller then showed that if T is any finite subset of $\{q_n : n\in\mathbb{N}\}$, then there is a continuous homomorphism $\varphi:\beta S_\omega\to S_\omega^*$ such that $\varphi[\beta S_\omega]=T$. Thus, unlike the situation in $\mathbb{N}$, continuous homomorphisms with finite images of any size are now known to exist. She also showed that this result does not apply to an infinite subset of $\{q_n : n\in\mathbb{N}\}$.

Dr. Miller also established that when the sequence $\langle q_n\rangle_{n=1}^\infty$ is constructed in Theorem 4.2, having chosen $\langle q_t\rangle_{t=1}^n$, there are $2^{\mathfrak{c}}$ choices for q_{n+1}, each in $K(\beta S_{n+1})$ and each less than q_n.

Monique Peters. We said at the start that we take βS to be a right topological with the additional property that for each $x\in S$, the function $q\mapsto x\cdot q$ is continuous. We have denoted the extended operation by the same symbol as used for the operation on S. We could alternatively have taken βS to be left topological. (In fact, that used to be my own choice.) Let us denote the left topological extension by $\odot$ to distinguish it from the right topological extension. Then one has that for each $p\in\beta S$, the function $q\mapsto p\odot q$ from βS to itself is continuous and for $x\in S$, the function $q\mapsto q\odot x$ is continuous.

If the operation on S is commutative, then $p\odot q=q\cdot p$ so $K(\beta S,\cdot)=K(\beta S,\odot)$. But if the operation is not commutative, there can be substantial differences. There are several notions of size that are relevant to the algebraic structure of βS, such as *syndetic*, *piecewise syndetic*, *thick*, and of course *central*. Each of these have both left and right versions, and in some cases it is easy to see that they are different. In my mind, the most difficult to tell apart are left and right J-sets.

Let us call the notion defined in Definition 4.1 a *right J-set*.

DEFINITION 4.3. Let $(S,\cdot)$ be a semigroup and let $\mathbb{F}$ be the set of sequences in S.

(a) Let $A \subseteq S$. Then A is a *left J-set* in S if and only if for each $F \in \mathcal{P}_f(\mathbb{F})$, there exist $m \in \mathbb{N}$, $a \in S^{m+1}$, and $t \in \mathcal{J}_m$ such that for each $f \in F$, $a(m+1)\cdot f\big(t(m)\big)\cdot a(m)\cdot f\big(t(m-1)\big)\cdot a(m-1)\cdots f\big(t(1)\big)\cdot a(1) \in A$.
(b) $(J(S,\odot) = \{p \in \beta S : (\forall A \in p)(A \text{ is a left } J\text{-set})\}$.

In her dissertation [**12**], Dr. Peters produced a subset A of the free semigroup S_ω on the distinct generators $\{a_n : n \in \mathbb{N}\}$ which is a left J-set but not a right J-set. The construction is far too complicated to be reproduced here. As a consequence, using the left-right switches of [**6**, Theorem 3.11 and Lemma 14.14.6] one also has that $J(S_\omega,\odot)\setminus J(S_\omega,\cdot) \neq \emptyset$.

5. Conclusion

In each of the seven dissertations that I have been discussing, there are a number of results that I haven't mentioned, and in some cases whole topics that I haven't hinted at. I have tried to present results that are, in my view, significant and also are reasonably easy to talk about without going into too much detail. The results of the seven dissertations extend our knowledge of Ramsey Theory and our understanding of the algebraic structure of the Stone-Čech compactification of a discrete semigroup. I want to thank all twenty of my Ph.D. students for helping to make my life interesting for over thirty years.

References

[1] Dibyendu De, Neil Hindman, and Dona Strauss, *A new and stronger central sets theorem*, Fund. Math. **199** (2008), no. 2, 155–175, DOI 10.4064/fm199-2-5. MR2410923

[2] H. Furstenberg, *Recurrence in ergodic theory and combinatorial number theory*, Princeton University Press, Princeton, N.J., 1981. M. B. Porter Lectures. MR603625

[3] A. W. Hales and R. I. Jewett, *Regularity and positional games*, Trans. Amer. Math. Soc. **106** (1963), 222–229, DOI 10.2307/1993764. MR143712

[4] Neil Hindman, *The research of thirteen students at Howard University*, Topology Appl. **156** (2009), no. 16, 2550–2559, DOI 10.1016/j.topol.2009.04.013. MR2561205

[5] Neil Hindman and Imre Leader, *Image partition regularity of matrices*, Combin. Probab. Comput. **2** (1993), no. 4, 437–463, DOI 10.1017/S0963548300000821. MR1264718

[6] Neil Hindman and Dona Strauss, *Algebra in the Stone-Čech compactification: Theory and applications; Second revised and extended edition [of MR1642231]*, De Gruyter Textbook, Walter de Gruyter & Co., Berlin, 2012. MR2893605

[7] Neil Hindman and Eric Tressler, *The first nontrivial Hales-Jewett number is four*, Ars Combin. **113** (2014), 385–390. MR3186481

[8] Henry Jordan, *Minimal Hales-Jewett Sets*, ProQuest LLC, Ann Arbor, MI, 2012. Thesis (Ph.D.)–Howard University. MR3004425

[9] John H. Johnson, *Some Differences Between an Ideal in the Stone-Cech Compactification of Commutative and Noncommutative Semigroups*, ProQuest LLC, Ann Arbor, MI, 2011. Thesis (Ph.D.)–Howard University. MR3004359

[10] Keith R. Milliken, *Ramsey's theorem with sums or unions*, J. Combinatorial Theory Ser. A **18** (1975), 276–290, DOI 10.1016/0097-3165(75)90039-4. MR373906

[11] K. Miller, *Continuous homomorphisms from βS to S^**, Ph.D. Dissertation, Howard University, 2013.

[12] Monique Agnes Peters, *Characterizing Differences between the Left and Right Operations on βS*, ProQuest LLC, Ann Arbor, MI, 2013. Thesis (Ph.D.)–Howard University. MR3218012

[13] Dev Raj Phulara, *A Generalization of the Central Sets Theorem With Applications and Some Additive and Multiplicative Ramsey Numbers*, ProQuest LLC, Ann Arbor, MI, 2014. Thesis (Ph.D.)–Howard University. MR3260015

[14] Kendra Enid Pleasant, *When Ramsey Meets Stone-Cech: Some New Results in Ramsey Theory*, ProQuest LLC, Ann Arbor, MI, 2017. Thesis (Ph.D.)–Howard University. MR3705979

[15] F. P. Ramsey, *On a Problem of Formal Logic*, Proc. London Math. Soc. (2) **30** (1929), no. 4, 264–286, DOI 10.1112/plms/s2-30.1.264. MR1576401

[16] Alan D. Taylor, *A canonical partition relation for finite subsets of* ω, J. Combinatorial Theory Ser. A **21** (1976), no. 2, 137–146, DOI 10.1016/0097-3165(76)90058-3. MR424571

[17] B. van der Waerden, *Beweis einer Baudetschen Vermutung*, Nieuw Arch. Wiskunde **19** (1927), 212-216.

[18] Kendall Williams, *Separating Milliken-Taylor systems and variations thereof in the dyadics and the Stone-Cech compactification of N*, ProQuest LLC, Ann Arbor, MI, 2010. Thesis (Ph.D.)–Howard University. MR2771544

Department of Mathematics, Howard University, Washington, DC 20059
Email address: nhindman@aol.com

Contemporary Mathematics
Volume **759**, 2020
https://doi.org/10.1090/conm/759/15272

Complex variables, mesh generation, and 3D web graphics: Research and technology behind the visualizations in the NIST Digital Library of Mathematical Functions

Bonita V. Saunders

Abstract. In 2010, the National Institute of Standards and Technology (NIST) launched the Digital Library of Mathematical Functions (DLMF), a free online resource containing definitions, recurrence relations, differential equations, and other crucial information about mathematical functions useful to researchers working in application areas in the mathematical and physical sciences. Although the DLMF was designed to replace the widely cited National Bureau of Standards (NBS) Handbook of Mathematical Functions commonly known as Abramowitz and Stegun (A&S), the goal was a compendium far beyond a book on the web, incorporating web tools and technologies for accessing, rendering, and searching math and graphics content. This paper focuses primarily on the research and technical challenges involved in creating the DLMF's graphics content, and in particular, its interactive 3D visualizations, where users can explore more than 200 graphs of high level mathematical function surfaces.

1. Introduction

In 2010, after a multi-year effort dating back to the late 1990s, the National Institute of Standards and Technology (NIST) released the Digital Library of Mathematical Functions (DLMF)[**9**], a free online resource containing definitions, recurrence relations, differential equations, and other crucial information to aid in the understanding and computation of mathematical functions that arise in application areas in the mathematical and physical sciences. Although the DLMF might be viewed as an update and replacement for the 1964 Handbook of Mathematical Functions with Formulas, Graphs, and Mathematical Tables (A&S) [**1**], in reality, it is quite different in both focus and content.

A&S was originally known for its tables. Its existence can be traced back to the Mathematical Tables Project created in 1938 by NIST's predecessor, the National Bureau of Standards (NBS), to address a crucial need for accurate tables to assist in the computation of functions commonly occuring in practical problems [**5**]. The project was administered by the Works Projects Administration, a New Deal agency of President Franklin Roosevelt. Highly educated, but out of work mathematicians and physicists supervised a staff of human 'computers' who performed calculations for reference tables of function values. From 1938 to 1946, 37 volumes of tables were published, including tables of trigonometric functions, logarithms,

2010 *Mathematics Subject Classification.* Primary 65D17.

the exponential function, and probability functions [**5**]. Realizing the importance of having the information all in one place, NBS mathematician Milton Abramowitz, a technical leader of the Mathematical Tables Project, eventually pushed for the publication of a compendium of tables and related material. This compendium, with emphasis on higher level functions such as Bessel functions, hypergeometric functions, and elliptic functions was published as the Handbook of Mathematical Functions with Formulas, Graphs, and Mathematical Tables in 1964. It is often simply called Abramowitz & Stegun or A&S in honor of its editors Milton Abramowitz and Irene Stegun who were Chief and Assistant Chief, respectively, of the Computation Laboratory of the NBS Applied Mathematics Division around the start of the project in 1956. Stegun took over the project and shepherded it to completion when Abramowitz died suddenly of a heart attack in 1958[**5**].

A casual glance at A&S clearly shows that tables predominate the handbook, but each function also includes related material such as formulas representing differential equations, definite and indefinite integrals, inequalities, recurrence relations, power series, asymptotic expansions, polynomial and rational approximations, graphs, and other qualitative information that might be useful for understanding and computing values of the function. For practicioners this "related material" has moved to the forefront over the years, while the tables have receded in importance due to the prevalence of reliable numerical software and computer algebra packages that have severely decreased the need for tables for computing function values by interpolation. Acknowledging this, it was decided that tables would not be included in the design for the DLMF. Nevertheless, the addition of new chapters on functions of growing significance and information on new properties of existing A&S functions increased the DLMF's content significantly over that of A&S.

The fact that the DLMF is web-based has opened up many possibilities that are still being explored. Navigational tools and hyperlinks allow the user to move around the site in a variety of ways. A database of metadata provides information for pop-up boxes where users can find links to sources, defined variables, cross references, alternative text formats such as LaTeX, MathML, and image formats, or notes on changes made to content. With the DLMF's math-aware search engine users can search by function names, particular formulas, and in some cases types of functions, for example, 'trig'. The site includes 600 2D and 3D plots along with 200 dynamic interactive visualizations where users can explore the graphs of elementary and high level mathematical functions. An overview of the technological capabilities in the DLMF was recently published in a Physics Today article[**20**]. In this paper we take a more in-depth view of the DLMF's graphics by looking at the ongoing research and development behind its interactive 3D visualizations.

Section 2 describes the techniques used to construct the computational grids for plotting the graphs shown in the visualizations. In Section 3 we look at our utilization and advancement of techniques for displaying interactive 3D graphics on the web and discuss some of the interesting capabilities available in the DLMF visualizations. Section 4 discusses ongoing challenges and future areas for research.

2. Grid Generation

During the design phase of the DLMF in the late 1990s and early 2000s little thought was given to how one might plot and render complex function graphs on

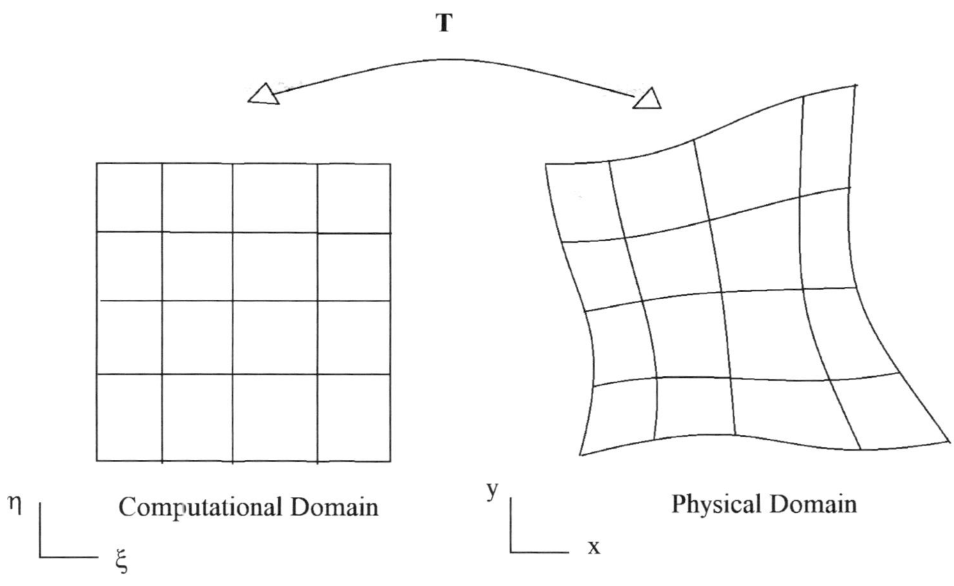

FIGURE 1. Numerical grid generation is defined by a curvilinear coordinate map from a canonical domain to the oddly shaped physical domain prescribed by the application.

the web. The focus was on the primary mathematical content that chapter authors would be asked to write. However, once a draft of the first chapter, on Airy and related functions, was written, we began looking at the best way to create the illustrative graphs the chapter needed. On looking at computer algebra systems we found that most did an excellent job plotting 2D graphs, which could be exported to an acceptable format for viewing on the web. For 3D graphs of function surfaces, the story was quite different. Most systems had rudimentary or non-existent machinery for properly clipping the graph when it was necessary to restrict the displayed range to illuminate significant features, such as poles or zeros. In one case the surface was properly clipped when viewed inside the system, but the unclipped data reappeared when the plot data was exported to a file. The problems we observed led us to design our own software for grid generation and, as we show in the next section, inspired our development of visualizations utilizing emerging web 3D technology.

We solved the clipping problem by computing the function over a 2D grid whose boundary included a level curve, or contour, of the function. We created the grid by using numerical grid generation, which defines a curvilinear coordinate system through a map from a canonical domain, such as a square in 2D, to the physical domain of interest, as shown in Figure 1.

Numerical grid generation is just one of several methods for creating a grid, or mesh, for solving problems over an oddly shaped domain. It has often been used with finite difference methods to solve partial differential equations (PDEs) governing flow around interesting geometries such as an airplane wing, ship's hull, or automobile body. Numerical grid generation is sometimes called structured grid generation because of the natural array order of the grid points on the physical domain [**23**]. Unstructured methods such as Delaunay triangulations, quadtree methods, or hybrid methods that combine both structured and unstructured meshes are often the methods of choice for extremely complex geometries. However, they require the storage of grid node connectivity information that can sometimes cause memory issues.

Structured methods can require a bit of ingenuity if the boundary shape is complex, but they may allow one to write more efficient code for some applications. In our case the structured order facilitated the coding of the interactive features for our visualizations. Our code is based on an algorithm we initially designed for problems in aerodynamics and solidification theory [**13, 14, 16**]. We modified the code to accurately approximate function boundary and contour data as well as capture significant function attributes such as poles, zeros, branch cuts, and other singularities.

When numerical grid generation is used for solving PDEs, the coordinate mapping must be one-to-one and onto to ensure invertibility. The goal is to transform the equations from the physical domain to equations over a simpler canonical domain where the difference equations and boundary conditions are easier to apply. Although, in our case, the 2D grid over the physical domain becomes the computational grid for the function being plotted, the same one-to-one correspondence is still needed since it ultimately affects the accuracy of the surface clipping and the smoothness of the colormap when the surface is rendered on the web [**15, 17**]

Our basic algorithm constructs a curvilinear coordinate spline mapping $\mathbf{T}$ from the unit square I_2 to the physical domain and is defined by

$$\mathbf{T}(\xi,\eta) = \begin{pmatrix} x(\xi,\eta) \\ y(\xi,\eta) \end{pmatrix} = \begin{pmatrix} \sum_{i=1}^{m}\sum_{j=1}^{n} \alpha_{ij} B_{ij}(\xi,\eta) \\ \sum_{i=1}^{m}\sum_{j=1}^{n} \beta_{ij} B_{ij}(\xi,\eta) \end{pmatrix} \tag{2.1}$$

where each B_{ij} is the tensor product of cubic B-splines. Therefore, $B_{ij}(\xi,\eta) = B_i(\xi)B_j(\eta)$ where B_i and B_j are elements of cubic B-spline sequences associated with finite nondecreasing knot sequences, say, $\{p_i\}_1^{m+4}$ and $\{q_j\}_1^{n+4}$, respectively [**13**].

To quickly obtain initial α_{ij} and β_{ij} coefficients for $\mathbf{T}$ we construct a transfinite blending function mapping [**10, 11, 23**] that interpolates the boundary of the physical domain. Conveniently, the spline coefficients can be divided into boundary coefficients that map the boundary of the square onto the boundary of the physical domain, and interior coefficients [**13, 14**], which hopefully map the interior of the square onto the interior of the physical domain. Their initial values are obtained by evaluating the transfinite interpolant at knot averages as described in [**4**], to produce a shape preserving approximation that reproduces straight lines and preserves convexity. If more accuracy is needed on part of the boundary, we use de Boor's SPLINT routine [**4**] to find coefficients that produce a cubic spline interpolant of that side. It is important that the spline knots and boundary coefficients be chosen carefully to produce an accurate representation of the physical boundary.

For simple boundaries, the initial coefficients produce a grid that is adequate for most applications, but if the boundary is more complicated or highly nonconvex, modifications of the coefficients are often necessary. To improve the grid, we fix the boundary coefficients and use a variational method to find interior coefficients that minimize the functional

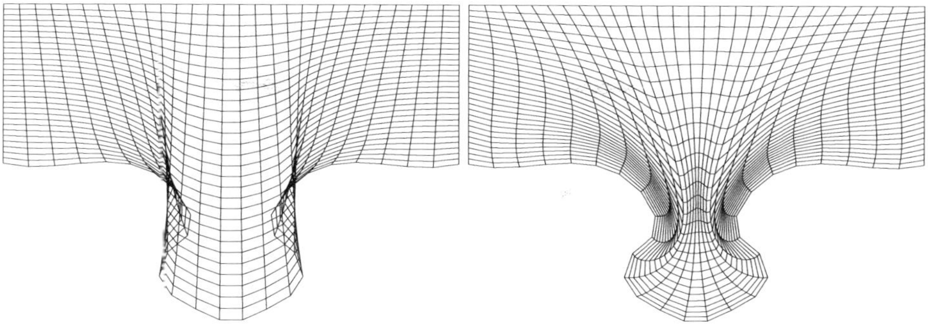

FIGURE 2. Initial and optimized puzzle grids.

$$F = \int_{I_2} \left(w_1 \left\{ \left(\frac{\partial J}{\partial \xi} \right)^2 + \left(\frac{\partial J}{\partial \eta} \right)^2 \right\} + w_2 \left\{ \frac{\partial \mathbf{T}}{\partial \xi} \cdot \frac{\partial \mathbf{T}}{\partial \eta} \right\}^2 \right) dA \tag{2.2}$$

where $\mathbf{T}$ denotes the grid generation mapping, J is the Jacobian of the mapping, and w_1 and w_2 are weight constants. This integral controls mesh smoothness and orthogonality. A large value for w_1 will decrease the variance in Jacobian values at nearby points, making the grid smoother. The w_2 weighted term represents the dot product of the tangent vectors $\partial\mathbf{T}/\partial\xi$ and $\partial\mathbf{T}/\partial\eta$. Therefore, minimizing this term enhances grid orthogonality. A change of variables shows this term to be equivalent to the volume weighted version of the orthogonality term in the Brackbill and Saltzman functional [**6**]. Figure 2 shows the initial and optimized grids for a physical domain shaped like a puzzle piece.

Figure 3 shows the computational grid and Riemann zeta function surface plot created using it. The grid boundaries, including the exterior boundary and the interior one around the pole, contain contour data for a height of 3. Computing the function over the grid produces a smooth clipping of the surface. A number of the "non-trivial" zeros of the Riemann zeta function can be viewed by exploring the figure on the DLMF site [**9**]. In our original code we input the location of the zeros to guarantee that there are gridpoints there. We are currently working on an algorithm that will automatically move gridpoints to the vicinity of a zero.

The current algorithm contains two significant changes over the original. First, we have added an adaptive term $w_3\{uJ^2\}$ to the integrand of the functional, where w_3 is a weight constant, and u represents external criteria for adapting the grid. If we were solving a system of partial differential equations, u might represent the gradient of the evolving solution or an approximation of truncation error. For our purposes, we want u to contain curvature and gradient information related to the function surface. The goal is to create a grid generation system that adaptively moves gridpoints to areas of high curvature or large gradient. With a change of variables this term is equivalent to the weighted volume variation, or adaptive, component of the Brackbill and Saltzman functional [**6, 23**]. Therefore, the enhanced integral should allow some control over mesh smoothness, orthogonality, and through u, permit an adaptive concentration of grid lines.

Second, a more fundamental change in our algorithm is replacing the mapping $\mathbf{T}$ by a composite mapping $\mathbf{T}^* = \mathbf{T} \circ \mathbf{\Phi}$ where $\mathbf{\Phi}$ is a tensor product spline mapping

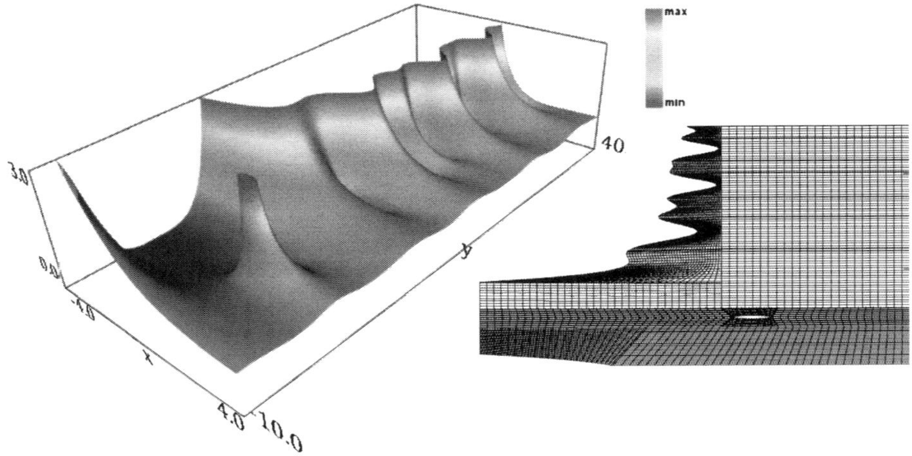

FIGURE 3. Riemann zeta function surface obtained by computing function over grid shown.

from the unit square I_2 to I_2 with its own coefficients and knot sequences [**18**]. Adaptive methods typically construct a reference grid [**22**] or distribution mesh [**13, 21**] by moving points on the canonical domain based on some adaptive criteria. The reference/distribution mesh is then mapped to the physical domain to create an adaptive mesh there. Mathematically this could be viewed as the composition of two maps where one maps the canonical domain to itself and the other maps the canonical domain to the physical domain. DeRose, et al., created composite maps in B-spline or Bézier form [**7, 8**].

For now, we have not tried to create a simple representation for our composite map, that is, we leave $\mathbf{\Phi}$ and $\mathbf{T}$ in their separate forms. The boundary coefficients of $\mathbf{T}$ can remain fixed while the coefficients of $\mathbf{\Phi}$ are adjusted to reparameterize the boundary points. The $\mathbf{\Phi}$ map can be used to create a reference grid that produces the desired adaptive effect without disturbing the accuracy of the boundary approximation.

After choosing initial $\mathbf{T}$ and $\mathbf{\Phi}$ coefficients that approximate transfinite interpolation we can improve our final physical grid, that is, the smoothness, orthogonality, or concentration of grid points by minimizing the following functional with respect to either the $\mathbf{\Phi}$ coefficients or interior $\mathbf{T}$ coefficients:

(2.3)
$$F^* = \int_{I_2} \left(w_1 \left\{ \left(\frac{\partial J^*}{\partial \xi} \right)^2 + \left(\frac{\partial J^*}{\partial \eta} \right)^2 \right\} + w_2 \left\{ \frac{\partial \mathbf{T}^*}{\partial \xi} \cdot \frac{\partial \mathbf{T}^*}{\partial \eta} \right\}^2 + w_3 \{ u {J^*}^2 \} \right) dA$$

where $*$ has been added to indicate the terms are associated with the composite mapping $\mathbf{T}^*$. Then J^*, the Jacobian of $\mathbf{T}^*$ is the product of J and J_Φ where J is the Jacobian of $\mathbf{T}$ and J_Φ, the Jacobian of $\mathbf{\Phi}$. To simplify our notation, $*$ is not added to the weight constants or u. Figure 4 shows a puzzle shaped grid adapted to the vertical line $x = 5.5$. We first optimize with respect to the interior $\mathbf{T}$ coefficients to obtain an acceptable physical grid as shown in Figure 2. We then optimize with respect to the $\mathbf{\Phi}$ coefficients to adapt the grid to the line. Our adaptive function u

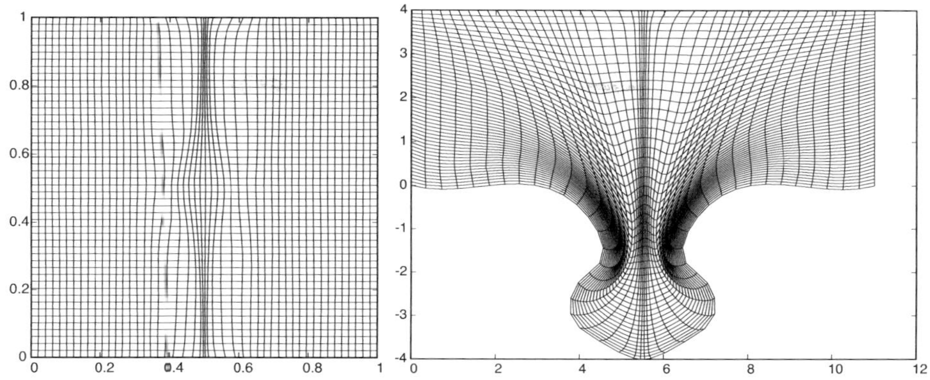

FIGURE 4. Reference grid and adapted puzzle grid.

is defined by

$$u(x, y) = e^{-50(x-5.5)^2}. \tag{2.4}$$

We have defined other expressions for u to adapt grids to circular arcs, circles, and intersecting lines [**18**]. We are now focusing on improving the performance of the code and experimenting with various definitions of u to capture function curvature and gradient data. Also, since our function data is computed using a variety of codes and computer algebra packages, we are also working on the integration of our grid generation code with various software packages and systems.

3. 3D web graphics and DLMF implementation

The individual chapters of both A&S and the DLMF were authored by various mathematicians and physicists of note. One A&S author was Philip J. Davis, who was at NBS at that time. Davis prepared the chapter on gamma and related functions, which he designed to serve as a model for the other authors. Davis hired Frank W.J. Olver who authored the A&S chapter on Bessel functions. Years later, Olver would serve as DLMF Mathematics Editor, Editor and Chief, and author several chapters in the DLMF.

More than 35 years after the publication of A&S, Davis, then a professor at Brown, was invited back to hear about NIST's plans for the development of the DLMF. Davis' tepid response to our preliminary colorful, but static 3D graphs for the first DLMF chapter, Airy and related functions, actually sparked our research and design of 3D function surface visualizations that grew in sophistication as technology for displaying 3D graphics on the web advanced.

3.1. 3D web graphics. As mentioned earlier, we found that at the start of the DLMF project in the late 1990s and 2000s, the export of 3D graphics data by well-known computer algebra systems was inadequate for our needs. After looking into the graphics technology being used and studied at NIST, we concluded that VRML (Virtual Reality Modeling Language) was our best option. VRML is a 3D file format for creating interactive graphics for viewing on the web. There were a variety of VRML viewer browsers that could be freely downloaded, but over time the maintenance of some of the best was discontinued and the quality of new browsers was mixed. Furthermore, as we began to better understand what features we wanted to see in DLMF visualizations, the complexity of our visualizations increased, making it more and more difficult to find VRML browsers that could handle our graphics files, even when our codes appeared to follow VRML standards.

Noting the industry transition from VRML to X3D (Extensible 3D), graphics team member Qiming Wang designed a VRML to X3D converter. By the launch of the DLMF in 2010 we had created close to 200 interactive visualizations of mathematical function surfaces accessible in both formats [**12**].

However, our ultimate goal was to make the DLMF visualizations accessible on Windows, Mac, and Linux platforms. We found VRML/X3D browsers that worked for Windows and Mac, but never found a Linux browser that could successfully handle our graphics files. Also, having to download a viewer browser/plug-in to see the visualizations was a headache for both maintainers and users of the DLMF site. Problems arose whenever the browsers needed to be updated, or whenever there were changes to the platform operating system.

Motivated by these concerns, in mid 2011 we began monitoring the development of WebGL, a JavaScript API (application programming interface) for rendering graphics in a web browser without a viewer plug-in. Then, thanks to the work of Johannes Behr and colleagues [**2**] on X3DOM, which permitted the direct integration of X3D nodes into HTML content, we were able to make a crucial decision. We would convert all the DLMF visualizations to WebGL by using the X3DOM framework to build the application around our X3D codes. We were encouraged by our early success in creating a few initial visualizations that worked in a beta WebGL accessible Mozilla Firefox browser. We were also bolstered by preliminary X3DOM/WebGL work by NIST researcher Sandy Ressler and the work of Steven Birr, et al., on the LiverAnatomyExplorer WebGL Tool [**3**] . We began an intensive effort to convert all the DLMF 3D visualizations to WebGL and seamlessly integrate the displays into the HTML pages of the associated chapters. The new visualizations first appeared in DLMF Version 1.0.7 released on March 21, 2014.

Building our application using the X3DOM framework allowed us to reuse most of our X3D code to create the WebGL files. The most challenging work was recoding the dynamic displays and interactive features. After first creating stand alone WebGL files, we worked with NIST computer scientist Brian Antonishek and Bruce Miller, information architect of the DLMF website, to make the coding changes needed to integrate the visualizations into DLMF HTML files. We also made style changes to achieve a more polished look. Most importantly we successfully achieved our main goal: To reproduce or enhance the capabilities available in our VRML/X3D visualizations and provide additional capabilities where possible. WebGL is now the default format for viewing the DLMF visualizations and VRML/X3D files are being phased out [**19**].

3.2. DLMF graphics features. The best way to experience the DLMF visualizations is to go directly to the graphics sections found in most chapters and explore the capabilities available. The visualizations can now be viewed in most common web browsers on Windows, Mac, or Linux platforms. A few features are highlighted in this section.

Figure 5 shows the general display for our surface visualizations. The user may click on the figure and rotate it freely or choose a stored viewpoint from the selection offered on the panel to the right. If a surface represents a complex valued

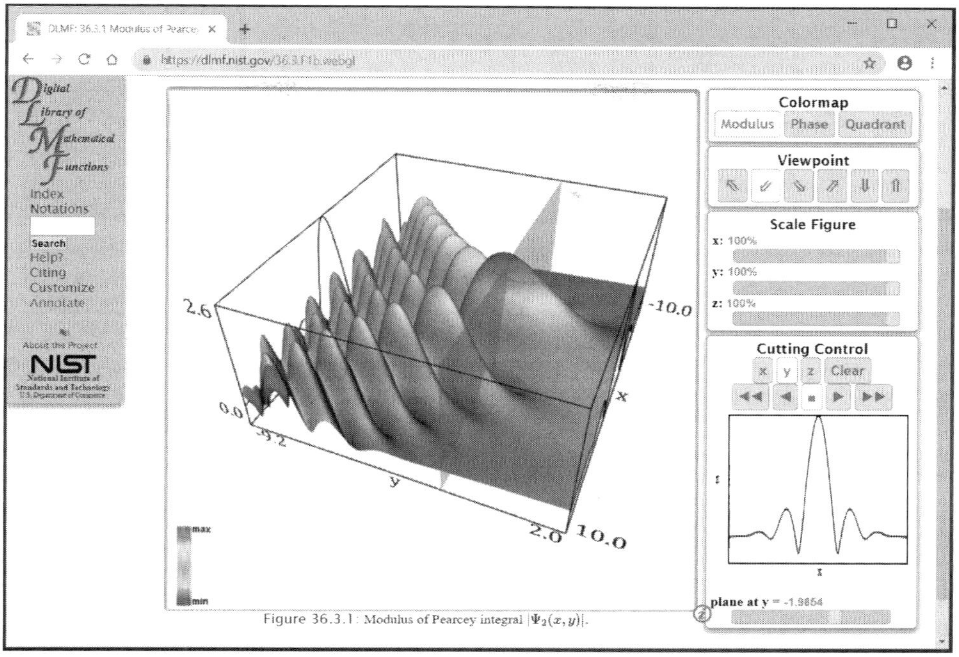

FIGURE 5. Modulus of Pearcey integral visualization embedded in DLMF webpage. Intersection of y direction cutting plane with surface displayed on bounding box sides and in pop-up display on side panel.

function, the user is offered a phase, or argument, based color map in addition to the height, or modulus, color map. This option will not appear if the function is real valued.

Figure 6 shows a density plot for a Jacobian Elliptic function adjacent to its surface plot. A user can apply the scaling option to collapse the function in the vertical direction to obtain the density plot.

At the top of Figure 7 a type of Bessel function known as a Hankel function is shown with a height-based color map. Its branch cut is evident when one switches to a phase color map, and on scaling the surface height to zero, the phase density plot shown at the bottom emerges. On the website, one should note the difference in how the color spectrum is traversed as one travels around a pole versus a zero.

4. Current and future areas for research

Clearly, a significant amount of work is involved in the design, creation, and maintenance of the visualizations in the DLMF. Initially, much of the 3D graphics work was motivated by deficiencies we saw in available software and computer algebra systems at the time. The rendering of 3D plots has improved in many systems, and export options have expanded tremendously, but we still notice that the quality of the 3D data exported may not match what is seen on the screen.

Creating our own grids and visualizations gives us access to the data and routines that control our visualizations. This is helpful if we want to expand existing

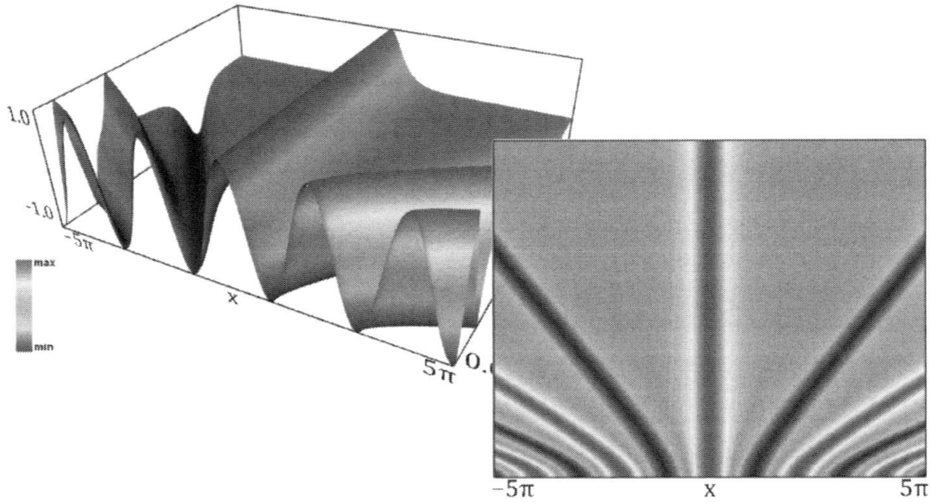

FIGURE 6. Jacobian Elliptic function $\mathrm{cn}(x, k)$ and density plot.

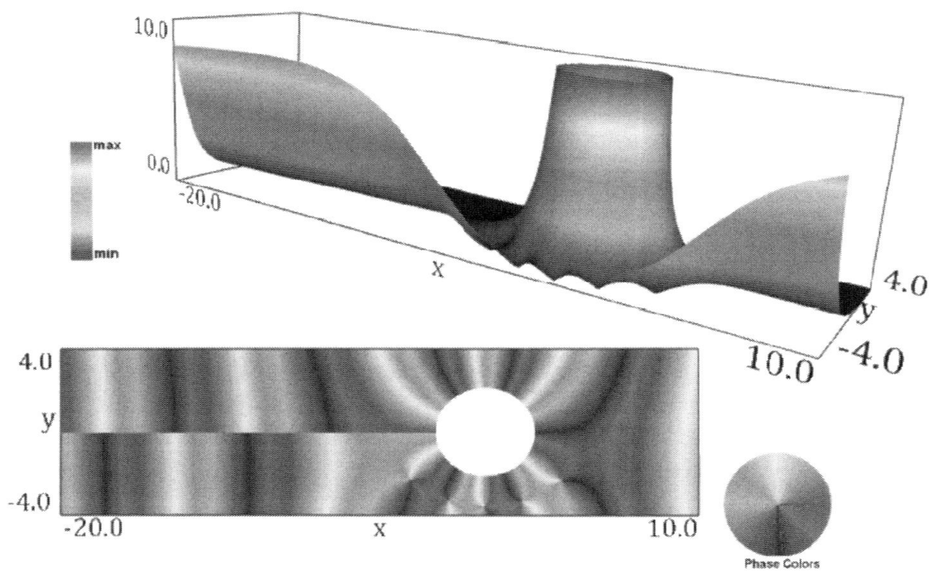

FIGURE 7. Modulus of Hankel function $H^{(1)}_{5.5}(x + iy)$ and phase density plot.

features or create new ones. Also, since our work is open to the public, we can get feedback from other researchers through publications and presentations at conferences.

There are several directions to go with our grid generation work. The ultimate goal is to develop a robust method that can be used to create quality grids in a reasonable amount of time. For now we will continue the work on adaptive curvature/gradient grids. We may also explore a parametric grid generation mapping which might work better if there are poles or other areas where there are steep gradients. Also, there have been some initial discussions with other grid generation researchers on the feasibility of creating a true zoom where the grid is refined and function values recomputed. Such an implementation would require a fast grid generation algorithm and hierarchical or locally refined techniques. Also, exploring unstructured triangulations and hybrid methods are still a possibility.

In addition to the true zoom, we might consider other changes to the visualizations such as adding or improving color maps, or including plots of real and imaginary parts of functions along with the modulus. In any case, while we expect to stick with our X3DOM/WebGL platform for the near future, we will strive to stay informed about trends in 3D web technology that might enhance our visualizations.

Disclaimer

All references to commercial products are provided only for clarification of the results presented. Their identification does not imply recommendation or endorsement by NIST.

References

[1] Milton Abramowitz and Irene A. Stegun, *Handbook of mathematical functions with formulas, graphs, and mathematical tables*, National Bureau of Standards Applied Mathematics Series, vol. 55, For sale by the Superintendent of Documents, U.S. Government Printing Office, Washington, D.C., 1964. MR0167642

[2] J. Behr, P. Eschler, and M. Zollner, X3DOM: A DOM-based HTML5/X3D Integration Model, in *Proceedings of the 14th International Conference on 3D Web Technology(Web3D '09)*, ACM, S.N. Spencer, ed., pp. 127-135, 2009.

[3] S. Birr, J. Monch,D. Sommerfeld, U. Preim, and B. Preim, The Liver Anatomy Explorer: A WebGL-based Surgical Teacing Tool, *IEEE Computer Graphics and Applications*, **33**, 5, pp.48-58, 2013.

[4] Carl de Boor, *A practical guide to splines*, Revised edition, Applied Mathematical Sciences, vol. 27, Springer-Verlag, New York, 2001. MR1900298

[5] R.F. Boisvert, D.W. Lozier, Handbook of Mathematical Functions, in *A Century of Excellence in Measurements Standards and Technology*(D.R. Lide, ed.), CRC Press, pp. 135-139, 2001.

[6] J. U. Brackbill and J. S. Saltzman, *Adaptive zoning for singular problems in two dimensions*, J. Comput. Phys. **46** (1982), no. 3, 342–368, DOI 10.1016/0021-9991(82)90020-1. MR673707

[7] T. D. DeRose, Composing Bezier simplexes, *ACM Transactions on Graphics* (3) **7** (1988), 198-221.

[8] T. D. DeRose, R. N. Goldman, H. Hagen & S. Mann, Functional composition algorithms via blossoming, *ACM Transactions on Graphics* (1993), 113-135.

[9] *NIST Digital Library of Mathematical Functions*. `https://dlmf.nist.gov/`, Release 1.022 of 2019-03-15. F.W J. Olver, A.B. Olde Daalhuis, D.W. Lozier, B.I. Schneider, R.F. Boisvert, C.W. Clark, B.R. Miller, and B.V. Saunders, eds.

[10] D. Gonsor, T. Grandine, A curve blending algorithm suitable for grid generation, in *Geometric Modeling and Computing: Seattle 2003*, Nashboro Press, Brentwood, 2004.

[11] William J. Gordon and Charles A. Hall, *Construction of curvilinear co-ordinate systems and applications to mesh generation*, Internat. J. Numer. Methods Engrg. **7** (1973), 461–477, DOI 10.1002/nme.1620070405. MR451775

[12] F.W.J. Olver, D.W. Lozier, R.F. Boisvert, C.W. Clark, A Special Functions Handbook for the Digital Age, *Notices Amer. Math Soc.* 58, 7, pp. 905-911, 2011.

[13] Bonita Valerie Saunders, *Algebraic grid generation using tensor product B-splines (computational fluids)*, ProQuest LLC, Ann Arbor, MI, 1985. Thesis (Ph.D.)–Old Dominion University. MR2634446

[14] B. V. Saunders, *A boundary conforming grid generation system for interface tracking*, Comput. Math. Appl. **29** (1995), no. 10, 1–17, DOI 10.1016/0898-1221(95)00040-6. MR1325278

[15] B. V. Saunders, Q. Wang, From 2d to 3d: numerical grid generation and the visualization of complex surfaces, in *Proceedings of the 7th International Conference on Numerical Grid Generation in Computational Field Simulations*, 51-60, Whistler, British Columbia, Canada, 2000.

[16] Bonita V. Saunders, *The application of numerical grid generation to problems in computational fluid dynamics*, Council for African American Researchers in the Mathematical Sciences, Vol. III (Baltimore, MD, 1997/Ann Arbor, MI, 1999), Contemp. Math., vol. 275, Amer. Math. Soc., Providence, RI, 2001, pp. 95–106, DOI 10.1090/conm/275/04492. MR1827337

[17] Bonita Saunders and Qiming Wang, *From B-spline mesh generation to effective visualizations for the NIST Digital Library of Mathematical Functions*, Curve and surface design: Avignon 2006, Mod. Methods Math., Nashboro Press, Brentwood, TN, 2007, pp. 235–243, DOI 10.6028/NIST.IR.7402. MR2335143

[18] B.V. Saunders, Q. Wang, B. Antonishek, Adaptive Composite B-Spline Grid Generation for Interactive 3D Visualizations, in *Proceedings of MASCOT12/ISGG2012(International IMACS Workshop and Bi-annual International Society for Grid Generation Conference)*, Las Palmas de Gran Canarias, 2012, IMACS Series in Computational and Applied Mathematics (Special Volume), 2014.

[19] B. Saunders, B. Antonishek, Q. Wang, B. Miller, Dynamic 3d visualizations of complex function surfaces using X3DOM and WebGL, in *Proceedings of the 20th International Conference on 3D Web Technology (Web3D 2015)*, Crete, Greece, 219-225, ACM, New York, 2015.

[20] B. Schneider, B. Miller, B. Saunders, NIST's Digital Library of Mathematical Functions, *Physics Today*, **71**(2):48–53, 2018, `https://doi.org/10.1063/PT.3.3846`.

[21] B. K. Soni, Grid generation for internal flow configurations, *Computers Math. Applic.* **24** (1992), 191-201.

[22] Stanly Steinberg and Patrick J. Roache, *Variational grid generation*, Numer. Methods Partial Differential Equations **2** (1986), no. 1, 71–96, DOI 10.1002/num.1690020107. MR925370

[23] Joe F. Thompson, Z. U. A. Warsi, and C. Wayne Mastin, *Numerical grid generation: Foundations and applications*, North-Holland Publishing Co., New York, 1985. MR791004

National Institute of Standards and Technology, 100 Bureau Drive, Stop 8910, Gaithersburg, Maryland 20899

Email address: bonita.saunders@nist.gov

Contemporary Mathematics
Volume **759**, 2020
https://doi.org/10.1090/conm/759/15273

A linear programming method for exponential domination

Michael Dairyko and Michael Young

ABSTRACT. For a graph G, the set $D \subseteq V(G)$ is a porous exponential dominating set if $1 \leq \sum_{d \in D} 2^{1-\text{dist}(d,v)}$ for every $v \in V(G)$, where $\text{dist}(d, v)$ denotes the length of the shortest dv path. The porous exponential dominating number of G, denoted $\gamma_e^*(G)$, is the minimum cardinality of a porous exponential dominating set. For any graph G, a technique is derived to determine a lower bound for $\gamma_e^*(G)$. Specifically for a grid graph H, linear programing is used to sharpen bound found through the lower bound technique. Lower and upper bounds are determined for the porous exponential domination number of the King Grid $\mathcal{K}_n$, the Slant Grid $\mathcal{S}_n$, and the n-dimensional hypercube Q_n.

1. Introduction

Domination in graphs is a tool used to model situations in which a vertex exerts influence on its neighboring vertices. For a graph G, a set $D \subseteq V(G)$ is a *dominating set* if every vertex contained in $V(G) \setminus D$ is adjacent to at least one vertex of D. The *domination number*, denoted $\gamma(G)$, is the cardinality of a minimum domination set.

Exponential domination was first introduced in [**6**] and is a variant of domination that models situations in which the influence an object exerts decreases exponentially as the distance increases. In particular, exponential domination models the dissemination of information in social networks where the information's influence decays exponentially with each share [**6**]. Therefore, exponential domination analyzes objects with a global influence. Other variants of domination investigate objects with local influence. There are two parameters within exponential domination; porous and non-porous. This paper focuses on porous exponential domination. A *porous exponential dominating set* is a set $D \subseteq V(G)$ such that $w^*(D, v) \geq 1$ for every $v \in V(G)$, where the weight function w^* is given by $w^*(u, v) = 2^{1-\text{dist}(u,v)}$ and $\text{dist}(u, v)$ represents the length of the shortest uv path. The *porous exponential domination number* of G, denoted by $\gamma_e^*(G)$, is the cardinality of a minimum porous exponential dominating set. For the sake of simplicity, we will refer to porous exponential domination as exponential domination. See Section 1.1 for technical definitions.

Section 2 develops a technique to determine the lower bound of the exponential domination number of any graph. Furthermore, with respect to grid graphs, a method using linear programing sharpens the lower bound. Section 3 applies

2010 *Mathematics Subject Classification.* Primary 05C69; Secondary 90C05.
Research supported by NSF Award #1719841.

the lower bound technique described in Section 2, to find lower bounds for the exponential domination number of the King grid $\mathcal{K}_n$, the Slant grid $\mathcal{S}_n$, and the n-dimensional hypercube Q_n. Upper bound constructions are then found for $\gamma_e^*(\mathcal{K}_n)$, $\gamma_e^*(\mathcal{S}_n)$ and $\gamma_e^*(Q_n)$.

1.1. Preliminaries. All graphs are simple and undirected. A graph $G = (V(G), E(G))$ is an ordered pair that is formed by a set of *vertices* $V(G)$ and a set of *edges* $E(G)$, where an edge is a two element subset of vertices. Consider a graph G and a set $D \subset V(G)$. Let $w : V(G) \times V(G) \to \mathbb{R}$ be a *weight function.* For $u, v \in V(G)$, we say that u assigns weight $w(u, v)$ to v. Denote the weight assigned by D to v as $w(D, v) := \sum_{d \in D} w(d, v)$, and similarly, the weight assigned by $d \in D$ to $H \subseteq V(G)$ as $w(d, H) := \sum_{h \in H} w(d, h)$. Let $\mathrm{m}(G) = \max_{d \in D} w(d, V(G))$. The pair (D, w) dominates G if $w(D, v) \geq 1$ for all $v \in V(G)$. The *excess weight* that the vertex v receives from D is defined as $\mathrm{exc}(D, v) = w(D, v) - 1$. We denote $\mathrm{exc}(D) = \sum_{v \in V(G)} \mathrm{exc}(D, v)$ to be the total excess weight that D sends out. Let $S_k(v) = \{u \in V(G) : \mathrm{dist}(u, v) = k\}$ denote the sphere of radius k.

Linear programing is an optimization technique that takes a set of linear inequalities, or constraints, and finds the best solution of a linear objective function. An integer program is a linear program, with the restriction that the variables can only be assigned integer values. Observe that $\gamma_e^*(G)$ is equivalent to finding the optimal value of the following integer program introduced by Henning et al:

Integer Program 1.1. [**9**]

$$\min \sum_{u \in V(G)} x(u)$$

$$\begin{aligned} s.t. \sum_{u \in V(G)} \left(\frac{1}{2}\right)^{\mathrm{dist}(u,v)-1} x(u) &\geq 1 \quad \forall v \in V(G) \\ x(u) &\in \{0,1\} \quad \forall u \in V(G). \end{aligned}$$

Notice that it is only feasible to run the program for graphs of small size, as the computation time for this integer program greatly increases as the size of the graph increases. To be able to run the program on graphs with larger sizes, the constraints in Integer Program 1.1 can be relaxed as shown in the following linear program.

Linear Program 1.2. [**9**]

$$\min \sum_{u \in V(G)} x(u)$$

$$\begin{aligned} s.t. \sum_{u \in V(G)} \left(\frac{1}{2}\right)^{\mathrm{dist}(u,v)-1} x(u) &\geq 1 \quad \forall v \in V(G) \\ x(u) &\geq 0 \quad \forall u \in V(G). \end{aligned}$$

The Cartesian product of two graphs G and H, denoted $G \square H$, is a graph such that $V(G \square H) = V(G) \times V(H)$ and two vertices $(g, h), (g', h')$ form an edge in $G \square H$ if and only if either $g = g'$ and h, h' form an edge in H, or $h = h'$ and g, g' form an edge in G. Let $G_{m,n} = P_m \square P_n$ be the *standard grid.* A *grid graph* is the standard grid with possibly additional edges added in a regular pattern. Notice that linear programming is a natural technique to apply to grid graphs. Observe that asymptotically, $G_{m,n}$ is equivalent to the torus $C_m \Box C_n$, which yields the same lower bound for the corresponding exponential domination number.

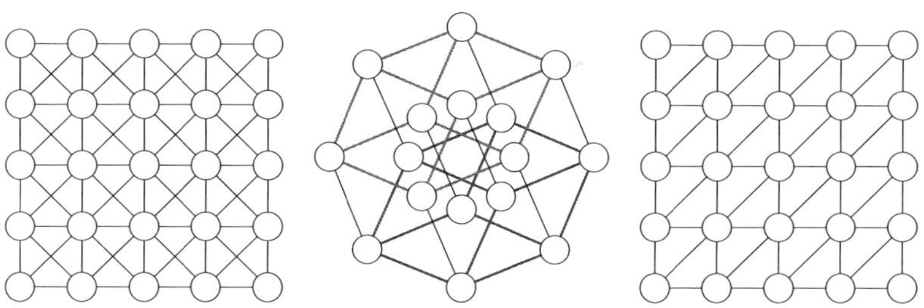

FIGURE 1. An illustration of $\mathcal{K}_5$, Q_4, and $\mathcal{S}_5$.

The *strong product* of two graphs G and H is the graph $G \boxtimes H$ for which $V(G \boxtimes H) = V(G) \times V(H)$ and two distinct vertices are adjacent whenever in both coordinate places the vertices are adjacent or equal in the corresponding graph. The *King grid* is defined as $\mathcal{K}_n = P_n \boxtimes P_n$. Let $[n] = \{1, 2, \ldots, n\}$. Consider the paths P_n and P_m with vertex sets $[n]$ and $[m]$, respectively. Then the *Slant* grid is defined to be $\mathcal{S}_n = P_n \Box P_m$ with the additional edges $\{i, j\} \sim \{i+1, j+1\}$, for $i \in [n-1]$ and $j \in [m-1]$. Notice that $\mathcal{K}_n$ and $\mathcal{S}_n$ are both instances of grid graphs. The *n-dimensional hypercube* graph, denoted Q_n, is constructed by creating a vertex for each n-digit binary number. Edges are formed if two vertices differ by one digit in their binary representation. See Figure 1 for an illustration of $\mathcal{K}_5$, Q_4, and $\mathcal{S}_5$.

1.2. Motivation. For $m \leq n$ consider $C_m \Box C_n$, the *torus* graph. Exponential domination of $C_m \Box C_n$ was first studied in [**1**]. Figure 2 is a visual representation of $C_{13} \Box C_{13}$, where 'X' denotes a member of D, an exponential domination set. Observe that there is one member of D in every row and column, therefore giving an upper bound construction for $\gamma_e(C_m \Box C_n)$ when m and n are multiples of 13. The following theorem extends this idea to large graphs.

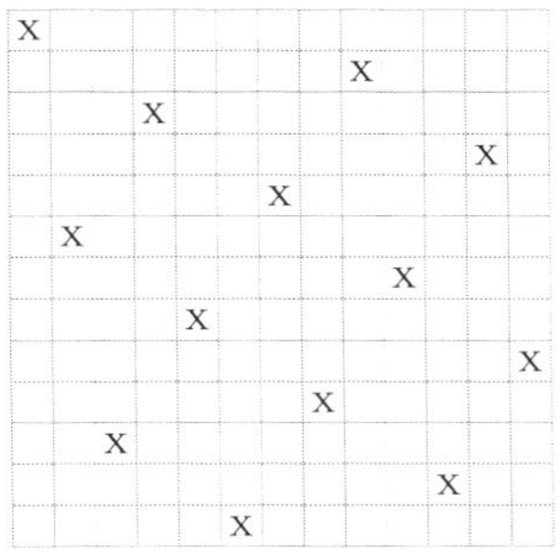

FIGURE 2. 13×13 Exponential Dominating Set Tile for $C_\infty \Box C_\infty$

THEOREM 1.3. [**1**] $\lim_{m,n \to \infty} \frac{\gamma_e^*(C_m \Box C_n)}{mn} \leq \frac{1}{13}$.

Notice that Theorem 1.3 directly implies that for $m, n \geq 13$, $\gamma_e^*(C_m \square C_n) \leq \left\lceil \frac{mn}{13} \right\rceil + o(1)$. Through a naive counting argument, it was shown that for $m, n \geq 3$, $\left\lceil \frac{mn}{15.875} \right\rceil \leq \gamma_e^*(C_m \square C_n)$ [**1**]. These results lead to the following conjecture.

CONJECTURE 1.4. *For all m and n, $\left\lceil \frac{mn}{13} \right\rceil \leq \gamma_e^*(C_m \square C_n)$.*

The lower bound for $\gamma_e^*(C_m \square C_n)$ was improved in [**5**] by taking the counting argument from [**1**] and applying it to linear programming.

THEOREM 1.5. [**5**] *For all $m, n \geq 11$, $\left\lceil \frac{mn}{13.761891939197298} \right\rceil \leq \gamma_e^*(C_m \square C_n)$.*

This paper was motivated by the work on determining $\gamma_e^*(C_m \square C_n)$ from [**1**] and [**5**]. The case specific lower bound technique from [**5**] is generalized to all graphs and the linear programming method detailed in [**5**] is generalized to all grid graphs.

2. A lower bound technique

In this section, a technique for determining the lower bound of the exponential domination number of any graph is derived. Through the use of linear programing, this technique is improved specifically for grid graphs. Note that the bound in Lemma 2.1 is sharp if $w^*(v, V(G)) = \mathrm{m}(G)$ for every $v \in V(G)$, where $m(G) = \max_{d \in D} w(d, V(G))$.

LEMMA 2.1. *Let D be an exponential dominating set for the graph G. If $k|D| \leq \mathrm{exc}(D)$, then*

$$\left\lceil \frac{|V(G)|}{\mathrm{m}(G) - k} \right\rceil \leq |D|.$$

PROOF. Observe that

$$\begin{aligned} |V(G)| &\leq \sum_{v \in V(G)} w(D, v) \\ &= \sum_{d \in D} \sum_{v \in V(G)} w(d, v) \leq |D| \, \mathrm{m}(G) - \mathrm{exc}(G) \leq |D| \left(\mathrm{m}(G) - k \right). \end{aligned}$$

□

REMARK 2.2. In Lemma 2.1, the value k is needed to compute the lower bound. For grid graphs, linear programming can be used to determine such a value of k. Mixed Integer Linear Program 2.3 is derived through the use of Linear Program 1.2 with two additional constraints. See Section 2.1 for the construction details. Let $x_{\min}$ be the optimal solution found from Mixed Integer Linear Program 2.3. As $w^*(D, v) \geq 1$ for all $v \in V(G)$, it follows that $|I| < x_{\min}$. Therefore $k = x_{\min} - |I|$.

MIXED INTEGER LINEAR PROGRAM 2.3.

$$\begin{array}{lrcl} \min & \sum_{i \in I} [A\mathbf{x}]_i & & \\ \text{s.t.} & A\mathbf{x} & \geq & \mathbf{1} \\ & A\mathbf{x} & \leq & b \\ & \mathbf{x} & \geq & \mathbf{0} \\ & x_i & \leq & 2, \; i \in I \\ & x_1 & = & 2. \end{array}$$

REMARK 2.4. Observe that Remark 2.2 localizes the global nature of exponential domination. Recall that exponential domination has a growth factor of $\frac{1}{2}$. This method can be applied to the variant of exponential domination with the growth factor of $\frac{1}{p}$ for $p \geq 3$. Furthermore, the method can be applied to other variants of domination to obtain a lower bound for the corresponding domination number. However, it is unclear whether the lower bound derived will be significant.

2.1. Mixed integer linear program setup. The setup for Mixed Integer Linear Program 2.3 is now discussed. Consider the $m \times n$ grid graph G and let D be a corresponding exponential dominating set. For a fixed $d_0 \in D$ and given an odd positive integer $r \leq \min\{m, n\}$, define H to be the $r \times r$ subgrid of G centered at d_0. Label the set of vertices $V(H)$ as $\{v_1, v_2, \ldots, v_{r^2}\}$ and let the indices of the interior vertices of H be defined as

$$I = \left\{i : v_i \in V(H) \text{ and } \operatorname{dist}(d_0, v_i) < \left\lfloor \frac{r}{2} \right\rfloor \right\}.$$

Then for $1 \leq k \leq r^2$, define $S'_k = v_k \cup \{u \in V(G \setminus H) : \operatorname{dist}(u, v_k) \leq \operatorname{dist}(u, h)\ \forall h \in V(H)\}$ and $x_k = w^*(S'_k \cap D, v_k)$. Notice that $S'_i = v_i$ for every $i \in I$. Therefore for $1 \leq k, j \leq r^2$, it follows that $w^*(S'_k \cap D, v_j) \leq x_k \left(\frac{1}{2}\right)^{\operatorname{dist}(v_k, v_j)}$. Thus, by the construction of S'_k,

$$w^*(D, v_j) \leq \sum_{k=1}^{r^2} w^*(S'_k \cap D, v_j) \leq \sum_{k=1}^{r^2} x_k \left(\frac{1}{2}\right)^{\operatorname{dist}(v_k, v_j)}.$$

Let A be the $r^2 \times r^2$ matrix such that $[A]_{kj} = \left(\frac{1}{2}\right)^{\operatorname{dist}(v_k, v_j)}$. Furthermore, let $\vec{x} = [x_1, x_2, \ldots, x_{r^2}]^\intercal$, where x_1 corresponds to d_0, and

$$\vec{w} = [w^*(D, v_1), w^*(D, v_2), \ldots, w^*(D, v_{r^2})]^\intercal.$$

Then observe that $\vec{w} \leq A\vec{x}$. The aim is to minimize $w^*(d_0, v_i)$ for all $i \in I$, while still satisfying that $w^*(D, v_i) \geq 1$. Therefore the objective function is to minimize $\sum_{i \in I} [A\mathbf{x}]_i$, where $\mathbf{x}$ is a vector of r^2 nonnegative variables.

Let $\mathbf{0}$ and $\mathbf{1}$ denote the $0s$ and $1s$ vectors of length r. Then the two constraints of Linear Program 1.2 with respect to the grid graph construction are that $A\mathbf{x} \geq \mathbf{1}$ and $\mathbf{x} \geq \mathbf{0}$. The remaining two constraints of Mixed Integer Linear Program are now discussed. By construction, any member of D assigns itself weight 2, and the remaining vertices do not have any initial weight. This gives the first integer constraint that $x_1 = 2$ and $x_i \leq 2$, for $i \in I$. Observe that it is necessary to determine an upper bound for $w^*(D, v_i)$ for each $v_i \in V(H)$ so that $w^*(d_0, v_i)$ can be decreased by the appropriate amount. To ensure this, we want

$$0 \leq w^*(d_0, v_i) - \operatorname{exc}(D, v_i) = w^*(d_0, v_i) - (w^*(D, v_i) - 1).$$

This implies that $w^*(D, v_i) \leq 1 + w^*(d_0, v_i)$. Let b be the real valued vector such that $b_i = 1 + \left(\frac{1}{2}\right)^{\operatorname{dist}(d_0, v_i) - 1}$ for $1 \leq i \leq r^2$. Therefore, the second constraint is $A\mathbf{x} \leq b$.

3. Main results

In this section the lower bound technique discussed in Section 2 is applied and upper bound constructions are found to bound the exponential domination number of the the King grid $\mathcal{K}_n$, Slant grid $\mathcal{S}_n$, and n-dimensional hypercube Q_n.

3.1. The King grid $\mathcal{K}_n$. Let D be an exponential dominating set for $\mathcal{K}_n$. Notice that for $d \in D$, it follows that $|S_k(v)| = 8k$ for $k \geq 1$. Then,

$$w^*(d, V(\mathcal{K}_n)) < 2 + \sum_{k=1}^{\infty} 8k \left(\frac{1}{2}\right)^{k-1} = 34.$$

This shows that $\text{m}(\mathcal{K}_n) < 34$. This fact, along with the optimal values of k determined by Mixed Integer Linear Program 2.3 can be applied with Lemma 2.1 to determine a lower bound for $\gamma_e^*(\mathcal{K}_n)$. See table 1 for a summary of these results. Observe that for $n \geq 11$, there is no feasible solution with Mixed Integer Linear Program 2.3. This is caused by the constraint $Ax \leq b$, since it puts a bound on the reduction of how much weight the center vertex can send out to the remaining interior vertices. Thus the best use of Mixed Integer Linear Program 2.3 will occur at $n = 7$.

TABLE 1. Lower Bounds for $\gamma_e^*(\mathcal{K}_n)$ for small values of n.

n	3	5	7	9	11
k	1	5.7806	10.6905	10.4103	$\emptyset$
$\gamma_e^*(\mathcal{K}_n) \geq$	$\frac{n^2}{33}$	$\frac{n^2}{28.2194}$	$\frac{n^2}{23.3095}$	$\frac{n^2}{23.5897}$	$\emptyset$

THEOREM 3.1. *For all $n \geq 7$, $\left\lceil \frac{n^2}{23.3095033018} \right\rceil \leq \gamma_e^*(\mathcal{K}_n)$.*

PROOF. Let D be a minimum exponential dominating set for $\mathcal{K}_n$. For each $d \in D$, let H be the 7×7 grid centered at d. The corresponding solution to Mixed Integer Linear Program 2.3 gives $x_{\min} = 35.6904966982$. Therefore let $k = 35.6904966982 - 25 = 10.6904966982$ and recall that $\text{m}(\mathcal{K}_n) < 34$. The result follows from Lemma 2.1. □

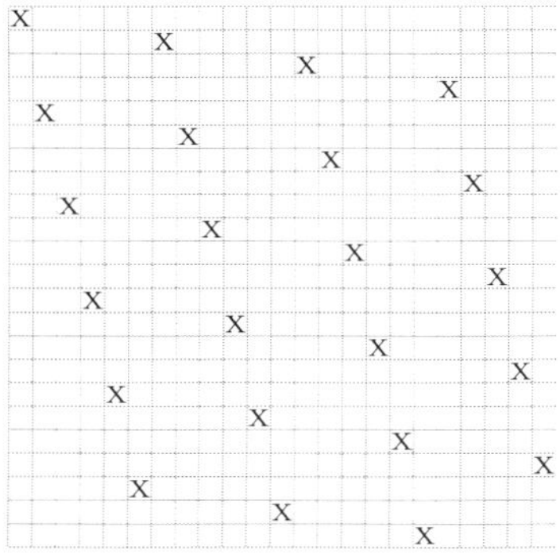

FIGURE 3. $T_{\mathcal{K}}$, the 23×23 exponential dominating set tile for $\mathcal{K}_\infty$.

Figure 3 shows a construction of a 23×23 tile $T_{\mathcal{K}}$, where 'X' denotes the location of an exponential dominating vertex. In particular, when $\mathcal{K}_\infty$ is tiled with $T_{\mathcal{K}}$, the exponential dominating set $D_{\mathcal{K}}$ is formed. The following theorem uses $T_{\mathcal{K}}$ to determines an upper bound for the asymptotic density of $\gamma_e^*(\mathcal{K}_n)$.

THEOREM 3.2. $\lim_{n\to\infty} \frac{\gamma_e^*(\mathcal{K}_n)}{n^2} \leq \frac{1}{23}$.

PROOF. Let $n = 23q + r$, for some $q, r \in \mathbb{Z}$ and $0 \leq r < 23$. Let H denote the $23q \times 23q$ subgrid of $\mathcal{K}_n$. Notice that we may tile H with the tiling scheme $T_{\mathcal{K}}$, as shown in Figure 3. Let $D_{\mathcal{K}}$ be the exponential dominating set that contains the $23q^2$ vertices used to tile H, as well as $V(\mathcal{K}_n \setminus H)$. Therefore $\gamma_e^*(\mathcal{K}_n) \leq 23q^2 + 46qr + r^2$, and we obtain the following asymptotic density:

$$\lim_{n\to\infty} \frac{\gamma_e^*(\mathcal{K}_n)}{n^2} \leq \lim_{q\to\infty} \frac{23q^2 + 46qr + r^2}{(23q + r)^2} \leq \frac{1}{23},$$

as the limit equals zero. □

THEOREM 3.3. *For all $n \geq 23$, $\gamma_e^*(\mathcal{K}_n) \leq \left\lceil \frac{n^2}{23} \right\rceil + o(1)$.*

PROOF. This result follows directly from Theorem 3.2. □

Similarly to Conjecture 1.4, we make the following conjecture.

CONJECTURE 3.4. *For all n, $\left\lceil \frac{n^2}{23} \right\rceil \leq \gamma_e^*(\mathcal{K}_n)$.*

3.2. The Slant grid $\mathcal{S}_n$. As the results in this section can be shown in a similar fashion as the results from Section 3.1, the proofs of Theorem 3.5 and Theorem 3.6, as well as other minor details will be omitted.

Let D be an exponential dominating set for $\mathcal{S}_n$. Notice that for $d \in D$, we have that $|S_k(d)| \leq 6k$ for $k \geq 1$. Then it follows that $\mathrm{m}(\mathcal{S}_n) < 26$.

This fact, along with the optimal values of k determined by Mixed Integer Linear Program 2.3 can be applied with Lemma 2.1 to determine a lower bound for $\gamma_e^*(\mathcal{S}_n)$. See table 2 for a summary of these results. Thus the best use of Mixed Integer Linear Program 2.3 will occur at $n = 7$.

TABLE 2. Lower Bounds for $\gamma_e^*(\mathcal{S}_n)$ for small values of n.

n	3	5	7	9
k	1.2353	3.9774	6.2655	$\emptyset$
$\gamma_e^*(\mathcal{S}_n) \geq$	$\frac{n^2}{24.7647}$	$\frac{n^2}{22.0226}$	$\frac{n^2}{19.7345}$	$\emptyset$

THEOREM 3.5. *For all $n \geq 7$, $\left\lceil \frac{n^2}{19.7344975348} \right\rceil \leq \gamma_e^*(\mathcal{S}_n)$.*

Figure 4 shows a construction of the 19×19 tile $T_{\mathcal{S}}$, such that when $\mathcal{S}_\infty$ is tiled with $T_{\mathcal{S}}$, the exponential dominating set $D_{\mathcal{S}}$ is formed. Notice that 'X' denotes the location of an exponential dominating vertex. The following theorem extends this tiling idea to determine an upper bound construction for $\gamma_e^*(\mathcal{S}_n)$.

THEOREM 3.6. *For $n \geq 19$, $\gamma_e^*(\mathcal{S}_n) \leq \left\lceil \frac{n^2}{19} \right\rceil + o(1)$.*

Similarly to Conjecture 1.4, we make the following conjecture.

CONJECTURE 3.7. *For all n, $\left\lceil \frac{n^2}{19} \right\rceil \leq \gamma_e^*(\mathcal{S}_n)$.*

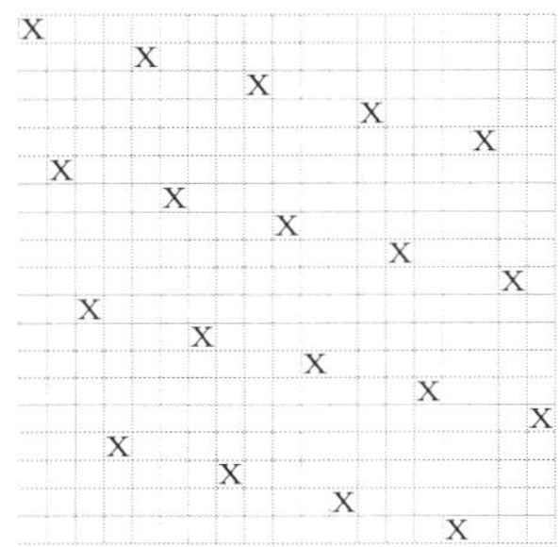

$T_{\mathcal{S}}$

FIGURE 4. $T_{\mathcal{S}}$, the 19×19 exponential dominating set tile for $\mathcal{S}_\infty$.

3.3. The n-dimensional hypercube. As Q_n is not a grid graph, the method used to determine a value of k for Lemma 2.1 in Remark 2.2 cannot be used to find the lower bound $\gamma_e^*(Q_n)$. In order to determine such a lower bound, a new method is used where distance properties of Q_n are exploited.

Let D be a minimum exponential dominating set for Q_n and let $d \in D$. Observe that for $u, v \in V(Q_n)$, the length of the shortest uv path in Q_n can be determined by the minimum number of digits that must be changed to get from u to v. Then for all $v \in V(Q_n)$,

$$w^*(v, V(Q_n)) = \sum_{i=0}^{n} \binom{n}{i} \left(\frac{1}{2}\right)^{i-1} = 2\sum_{i=0}^{n} \binom{n}{i} \left(\frac{1}{2}\right)^{i} \cdot 1^{n-i} = 2\left(\frac{1}{2}+1\right)^{n} = 2\left(\frac{3}{2}\right)^{n}.$$

Thus it follows that $\mathrm{m}(Q_n) = 2\left(\frac{3}{2}\right)^n$.

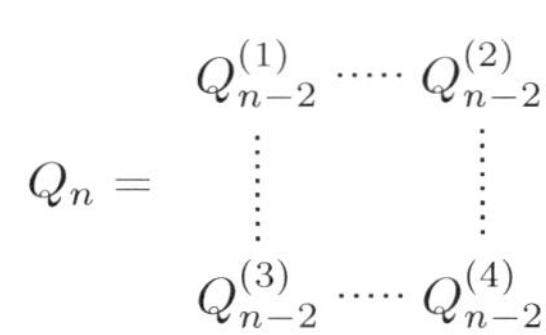

FIGURE 5. A decomposition of Q_n, where $Q_n = Q_{n-2} \square K_2 \square K_2$.

In the following theorem, the decomposition of Q_n in Figure 5 and value of $\mathrm{m}(Q_n)$ are used to show that $\left(\frac{4}{3}\right)^n \leq \gamma_e^*(Q_n) \leq (\sqrt{2})^n$ for large n.

THEOREM 3.8. *For all* $n \geq 1$, $\left\lceil \dfrac{2^{n+3}}{2^{4-n} \cdot 3^n - 2n - 9} \right\rceil \leq \gamma_e^*(Q_n) \leq (\sqrt{2})^n$.

PROOF. We begin with the lower bound. Let D be an exponential dominating set for Q_n and suppose, without loss of generality, that $d = \{0, 0, \ldots, 0\} \in D$. Let $A = \{a \in V(Q_n) : \ a \text{ has an odd number of } 1's\}$ and $B = V(Q_n) \setminus (A \cup \{d\})$. Let $X = \{x \in V(Q_n) : \ dx \in E(Q_n)\} \subset A$ and $Y = \{y \in V(Q_n) : \ xy \in E(Q_n) \text{ for some } x \in X\} \subset B$. Then $w^*(d, X) = |X| = n$ and $w^*(d, Y) = \frac{n}{2}$. As (D, w^*) dominates Q_n, $w^*(D \setminus d, Y) \geq \frac{n}{2}$. This implies that $w^*(D \setminus d, X) \geq \frac{n}{4}$, and $w^*(D \setminus d, d) \geq \frac{1}{8}$. Therefore $\mathrm{exc}(D, X) \geq \frac{n}{4}$ and $\mathrm{exc}(D, d) = \frac{9}{8}$, which holds for all

$d \in D$. This gives that

$$\operatorname{exc}(D) \geq \left(\frac{9}{8} + \frac{n}{4}\right)|D| = \frac{2n+9}{8}|D|.$$

Then using $\mathrm{m}(Q_n) = 2\left(\frac{3}{2}\right)^n$ and $k = \frac{2n+9}{8}$, the lower bound follows from Lemma 2.1.

Now we show the upper bound. From Figure 5, $Q_n = Q_{n-2} \square K_2 \square K_2$. Without loss of generality, let D and D' be two minimum exponential dominating sets for $Q^{(1)}_{n-2}$ and $Q^{(4)}_{n-2}$, respectively, with labeling as in Figure 5. Therefore it follows by definition that $w^*(D, V(Q^{(1)}_{n-2})) \geq 1$ and $w^*(D', V(Q^{(4)}_{n-2})) \geq 1$. As neighboring vertices also receive weight, $w^*(D, V(Q^{(2)}_{n-2}))$, $w^*(D, V(Q^{(3)}_{n-2})) \geq \frac{1}{2}$ and $w^*(D', V(Q^{(2)}_{n-2}))$, $w^*(D', V(Q^{(3)}_{n-2})) \geq \frac{1}{2}$. This implies that $D \cup D'$ forms an exponential dominating set for Q_n. Let $a_n = \gamma_e^*(Q_n)$ and $a_{n-2} = |D| = |D'|$, so $a_n = 2a_{n-2}$. We now show that $a_n \leq 2^{\frac{n}{2}}$ by induction. Observe that when $n = 2$, we have that $a_2 = 2 \leq 2^1$. Now suppose that $a_n \leq 2^{\frac{n}{2}}$ holds for all $n < k$. Now consider the case when $n = k$. Then using the inductive hypothesis,

$$a_k = 2a_{k-2} \leq 2(2^{\frac{1}{2}(k-2)}) = 2^{\frac{k}{2}}.$$

Therefore by induction, $\gamma_e^*(Q_n) \leq (\sqrt{2})^n$. $\square$

Acknowledgments

We would like to thank Dr. Leslie Hogben for her input on this paper.

References

[1] Mark Anderson, Robert C. Brigham, Julie R. Carrington, Richard P. Vitray, and Jay Yellen, *On exponential domination of* $C_m \times C_n$, AKCE Int. J. Graphs Comb. **6** (2009), no. 3, 341–351. MR2589713

[2] A. Aytaç and B. Atay, *On exponential domination of some graphs*, Nonlinear Dyn. Syst. Theory **16** (2016), no. 1, 12–19. MR3495635

[3] Stéphane Bessy, Pascal Ochem, and Dieter Rautenbach, *Exponential domination in subcubic graphs*, Electron. J. Combin. **23** (2016), no. 4, Paper 4.42, 17, DOI 10.37236/5711. MR3604800

[4] Stéphane Bessy, Pascal Ochem, and Dieter Rautenbach, *Bounds on the exponential domination number*, Discrete Math. **340** (2017), no. 3, 494–503, DOI 10.1016/j.disc.2016.08.024. MR3584835

[5] Chassidy Bozeman, Joshua Carlson, Michael Dairyko, Derek Young, and Michael Young. Lower bounds for the exponential domination number of $C_m \times C_n$. `arXiv:1803.01933`.

[6] Peter Dankelmann, David Day, David Erwin, Simon Mukwembi, and Henda Swart, *Domination with exponential decay*, Discrete Math. **309** (2009), no. 19, 5877–5883, DOI 10.1016/j.disc.2008.06.040. MR2551966

[7] Teresa W. Haynes, Stephen T. Hedetniemi, and Peter J. Slater, *Fundamentals of domination in graphs*, Monographs and Textbooks in Pure and Applied Mathematics, vol. 208, Marcel Dekker, Inc., New York, 1998. MR1605684

[8] Michael A. Henning, Simon Jäger, and Dieter Rautenbach, *Hereditary equality of domination and exponential domination*, Discuss. Math. Graph Theory **38** (2018), no. 1, 275–285, DOI 10.7151/dmgt.2006. MR3743966

[9] Michael A. Henning, Simon Jäger, and Dieter Rautenbach, *Relating domination, exponential domination, and porous exponential domination*, Discrete Optim. **23** (2017), 81–92, DOI 10.1016/j.disopt.2016.12.002. MR3610967

Milwaukee Brewers, Milwaukee, WI 53214
Email address: mike.dairyko@brewers.com

Iowa State University, Ames, IA 50011
Email address: myoung@iastate.edu

Contemporary Mathematics
Volume **759**, 2020
https://doi.org/10.1090/conm/759/15274

The first 50 years of the National Association of Mathematicians, Inc. (NAM)

Johnny L. Houston

ABSTRACT. This paper provides a general overview of the history of the National Association of Mathematicians (NAM) as a professional organization in the mathematical sciences. It begins with the Founding of NAM (1969) and culminates with NAM's 50th Anniversary Celebration (2019). Its aim is not to provide all the details of the first 50 years but rather to emphasize some of the major activities, contributions, and influences that NAM has made in the underrepresented American minority communities and in the larger Mathematical Sciences community. The initial goals of NAM's members were Awareness, Recognition, Inclusion, and Cooperation. However, the early members of NAM established a greater vision for NAM's future: creating a community of scholars, developing scholars for cultivating more scholars and the promotion of excellence in the Mathematical Sciences. Their vision also included establishing exemplary programs that encourage the development of scholars as well as recognize and honor past outstanding mathematicians of underrepresented American minorities. As NAM pauses to reflect during its Golden Anniversary, the organization is taking a bold new look at its future once again as it moves into the next fifty years.

This document presents NAM today (2019), with a historical perspective and a vision of its future. For those readers not familiar with NAM as an organization, this document is written to provide only a brief, but cumulative history of NAM in 2019 during NAM's Fiftieth (50th) Anniversary Year. It is not a detailed description encompassing all of NAM's activities over a fifty-year period but does provide several major highlights that have occurred during the history of the organization (1969-2019). This document addresses a simple question: What has characterized NAM's existence for the first fifty years? Succinctly stated, the first fifty years of NAM can be characterized as an era for visionary and transformational changes in the mathematical sciences community for the American minority and majority, nationally and across all levels of academia.

1. The establishment of NAM

A group of seventeen persons, all Black and Brown mathematical professionals, are given credit for bringing about the establishment of NAM. The group met in New Orleans Sunday, January 26, 1969, to discuss how to proceed as not only a group, but as a voice, and as a positive force that would make a difference in the

---

2010 *Mathematics Subject Classification.* Primary 01A61, 01A65, 01A70.

mathematical sciences community in the United States, and indeed, in the world. This force would be a clear and omnipresent voice for issues, ideas, perspectives, and for those who did not have access to such a voice in the past. They would advocate inclusion in the mathematical sciences community and more importantly, would refuse to become isolated from the mainstream. These mathematicians would further advocate conflict resolution and human/cultural problem-solving for the common good of the mathematical sciences community. The ultimate mission and purpose of NAM would become as follows:

- **To promote excellence in the mathematical sciences; and**
- **To promote the mathematical development of under-represented American minorities.**

Several records of the Association note what transpired; but most importantly, this was a meeting igniting a flame that resulted in the establishment of NAM. The seventeen mathematicians themselves were not so special or important individually but were of one accord in representing, the views and perspectives of American minorities in the mathematical sciences. Moreover, they committed themselves in ushering in the new era of mathematical sciences in the United States. They developed the resolve that silence and exclusion would no longer be the order of the day and instead, all who desired to learn mathematics and/or contribute to the community of scholars in the mathematical sciences, would be both encouraged and assisted in doing so. With this resolve the group raised the following questions:

If not us as spokespersons, then who?
If not now to begin change, then when?

Their resolve was that from that day forward, American mathematical professionals of color would embrace the Kwanzaa Principle of Kujichagulia (Self-Determination).

Thus, the era of NAM's first five decades began. In less than a year, these seventeen voices had been joined, literally, by scores of others hoping to accomplish the same goal. In less than two years, this force had become an established organization entitled:

The National Association of Mathematicians, Inc. (NAM)

FIGURE 1. Left: Rogers Newman, Johnny Houston, and Benjamin Martin; Right: Etta Falconer, Evelyn Granville, Lee Lorch, and Vivienne Malone-Mayes

TABLE 1. The Seventeen Founders of NAM

1.	James A. Donaldson	University of Illinois at Chicago
2.	Samuel Douglas	Grambling College
3.	Henry Eldridge	Fayetteville State College
4.	Thyrsa Frazier-Svager	Central State University
5.	Richard Griego	University of New Mexico / Albuquerque
6.	Johnny L. Houston	Stillman College / Grad Student at Purdue
7.	Curtis Jefferson	Cuyahoga Community College
8.	Vivienne Malone-Mayes	Baylor University
9.	Theodore Portis	Alabama State University
10.	Charles Smith	Paine College
11.	Robert Smith	Pennsylvania State University
12.	Beauregard Stubblefield	Texas Southern University
13.	Henry Taggert	Jarvis Christian College
14.	Walter Talbot	Morgan State College
15.	Argelia Valez-Rodriquez	Bishop College
16.	Harriet J. Walton	Morehouse College
17.	Scott Williams	Lehigh University

2. Commanding and providing recognition

As NAM continued to meet, define itself and set its agenda, other mathematical sciences organizations, governmental and private agencies, began to give recognition to NAM. Once it was understood that NAM's agenda was neither exclusive or negative in nature, that NAM communicated and negotiated with others in a professional manner and that when NAM disagreed (when necessary) it did so in an agreeable manner, many traditional mathematicians joined NAM in support of its mission and purpose. Other mainstream mathematicians, who did not join NAM, came to respect the organization. Over the years, respect for NAM has grown from small circles of mathematical professionals, to larger circles: regional, national, and international arenas.

Thus, in a short period of time, NAM made others more aware of its existence as well as its mission. This led to the change in the composition of National Boards and Committees as well as the Boards and Committees of other mathematical sciences organizations. NAM, collectively, and many men and women of color, individually, began receiving invitations to participate and to be involved in all levels of activities in the mathematical sciences community. Today NAM is considered a peer by most mathematical science organizations. For example, NAM is considered a junior partner in the Joint Winter Mathematics Meetings of the AMS/MAA, held each January and in MAA's Annual Summer MathFest. NAM is currently considered a full partner of the Conference Board of Mathematical Sciences (CBMS), a full partner/supporter of the American Mathematics Competitions and a full partner/supporter of IMO, the International Mathematics Olympiad. In short, NAM is currently recognized and respected, nationally and internationally, as a viable and contributing non-profit mathematical science organization of quality.

FIGURE 2. Early NAM Presidents: Johnny Houston (Acting President,1969), Frank James (NAM's first Elected President, 1970), Rogers Newman (President, 1984-1994),and John Alexander (President, 1994-2004)

3. The primary visionary goals of NAM:

A. To promote excellence in the mathematical sciences.

B. To promote the mathematical development of American minorities.

C. To exhibit and command, by example and by mutual respect, how all Americans in the mathematical sciences, both the majority and minority members should be acknowledged, respected, and valued on the national scene and at all levels of academia (and in all institutions) when they demonstrate noteworthy scholarship in the mathematical sciences.

D. To create and continuously cultivate a community of scholars in the mathematical sciences with membership open to all and with activities inclusive to other mathematical communities of scholars. NAM would also advocate cooperation and the practice of sharing among all mathematical communities.

E. To organize and maintain a state-of-the-art 501 (c) (3) professional organization.

F. To establish an organization with structure that will develop and cultivate exemplary programs and activities encouraging and challenging present scholars/developing scholars, as well as acknowledge and honor some past outstanding scholars who are worthy of being emulated.

G. To continuously seek new approaches to encourage more American minorities to pursue PhDs in the mathematical sciences; open, negotiate, and create more opportunities for American minorities to earn PhDs in the mathematical sciences so that the annual percentage of American minorities earning such degrees will gradually approach parity (to their percentage population-wise).

The realization of **Goals A** through **D** has been paramount since the establishment of NAM, **Goal E** was incorporated in the early 1970s, and lastly **Goals F and G** have been constantly evolving and can best be addressed by describing NAM's programs and annual activities.

4. The establishment of named annual lecture series

With the establishment of its named lecture series, NAM has given respect and recognition to the following African American mathematicians:

A. The **Claytor Lecture** was established in 1980 and is presented annually at the Joint Winter Mathematics Meetings in January in honor of **W. S. Claytor (1908-1967)**, the first nationally recognized African American research mathematician. Claytor was the third African American to receive a PhD degree in mathematics (1933). The lecture is given by an African American researcher in the mathematical sciences. Later the lecture was re-named the **Claytor-Woodard Lecture**. **Dudley Weldon Woodard** was Claytor's professor at Howard University and the second African American to earn a PhD in mathematics.

B. The **Blackwell Lecture** was established in 1994 and is presented at the MAA MATHFest (Summer Meeting) in honor of **David Blackwell (1919-2010)**, the only African American mathematician elected to the National Academy of Science (Fellow.)

C. The **Wilkins Lecture** was established in 1994 and is presented at NAM's Undergraduate MATHFest in the fall in honor of **J. Ernest Wilkins, Jr. (1923-2011)**. Wilkins is one of two African American mathematician-engineers elected as a Fellow to the National Academy of Engineers.

D. The **Bharucha-Reid Lecture** was established in 1994 and is presented at NAM's annual Faculty Conference on Research and Teaching Excellence (FCRTE) in the spring in honor of **Albert Turner Bharucha-Reid (1922-1985)**. Bharucha-Reid was a world-class mathematician who only earned a BS degree yet published 75 referred papers and wrote five books. In addition, at least thirteen students have earned a PhD in mathematics under his supervision as dissertation advisor.

In addition to honoring the mathematicians with the named lecture series, NAM uses these lectures each year to give recognition to persons who are currently engaged in research. It is NAM's belief that by extending an invitation to present their research, these mathematicians will be encouraged to continue to develop their research careers.

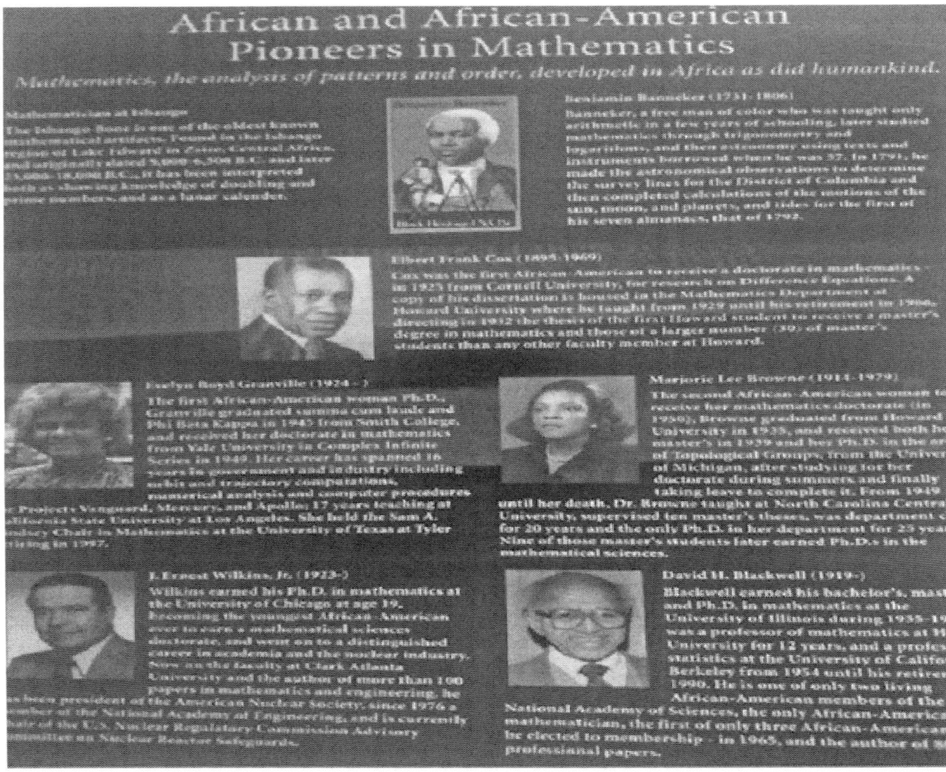

FIGURE 3. Top to Bottom: Benjamin Banneker, Elbert Cox, Evelyn Boyd Granville, Marjorie Lee Browne, J. Ernest Wilkins, and David H. Blackwell

5. Establishment of other named activities

With the establishment of the **Cox-Talbot Address** in 1980, NAM has given honor to two African American Mathematicians, **Elbert F. Cox (1895-1969)**, the first black man in the world to receive a PhD degree in mathematics and to **Walter R. Talbot (1909-1977)**, the elder statesman in the founding of NAM and the fourth black American to receive a PhD degree in mathematics. Each year, this address permits a mathematical professional to share pertinent information with the mathematical sciences community. Additionally, with the establishment of the **Granville-Browne Session for Presentations by Recent Recipients of the Doctoral Degree** (1996) the mathematical sciences, NAM gave honor to two of earlier African American women to receive the PhD degree in mathematics: **Evelyn Boyd Granville (1924 -)** who received her PhD in 1949, and **Marjorie Lee**

Browne (1914-1979) who received her degree in 1950. These women, by their impressive professional lives, have inspired and encouraged many young scholars to continue to pursue excellence in mathematical sciences. Even today Dr. Granville continues to inspire and most recently attended NAM's 50th Anniversary Celebration and received **NAM's Legacy Award**. Moreover, these two women join NAM (vicariously) in the encouragement of today's young men and women (who present in the Granville-Browne session at the Joint Winter Mathematics Meetings) to develop research careers and pursue professional lives in the mathematical sciences. Today it is known as the **Haynes-Granville-Browne Symposium** as it was discovered later that **Euphemia Lofton Haynes (1890-1980)** was the first black woman to receive a PhD in mathematics in 1943.

6. The establishment of various NAM awards: the 1990s

Lifetime Achievement Awards. This award honors distinguished mathematical professionals whose careers over a period of twenty-five years or longer have been exemplary and worthy of emulating. To date, NAM has presented its Lifetime Achievement Award to:

David Blackwell	(1994)	J. Ernest Wilkins	(1994)
Lee Lorch	(1995)	Evelyn Boyd Granville	(1996)
Charles Bell	(1997)	Clarence F. Stephens	(1998)
Johnny L. Houston	(1999)	Beauregard Stubblefield	(2000)
Evelyn Dawley Green	(2001)	John W. Alexander, Jr.	(2005)
Etta Z. Falconer	(2003 - P[1])	Robert E. Bozeman	(2011)
Abdulalim A. Shabazz	(2010)	Jacquelyn Giles	(2015)
Sylvia T. Bozeman	(2012)	Nagambal Shah	(2017)
William Hawkins	(2016)	Carolyn Mahoney	(2018)
Rudy Horne	(2018 - P)	Melvin Currie	(2019)

NAM's Centenarian Award.

Virginia Newell	(2018)	(born October 7, 1917)
Clarence F. Stephens	(2018)	(July 24, 1917 - March 5, 2018)
Katherine G. Johnson	(2019)	(born August 26, 1918)

NAM's Stephens-Shabazz Teaching Award.

Duane Cooper (2019)

Because NAM is essentially a volunteer-based professional organization in the mathematical sciences, most of NAM's success has been due to the contributions from hundreds of individuals who identify with NAM's mission helped in making a difference. To express appreciation to the many persons who have gone above and beyond the average in their contributions, NAM established and has presented **Distinguished Service Awards and Awards of Appreciation** to these men and women. For example, the Association presents an award to each person who gives a NAM lecture or address annually, to each major presenter at Undergraduate MATHFest, to board members who have served on NAM's Board of Directors for a designated minimum period of time, local coordinators of NAM activities, and to persons who have made significant contributions through NAM to the mathematical sciences community.

[1]P stands for posthumously.

FIGURE 4. Left: Clarence Stephens and Johnny Houston; Right: J. Ernest Wilkins, Johnny Houston, and David Blackwell

FIGURE 5. Left: NAM Regional Faculty Conference (FCRTE); Right: John Nash and Johnny Houston

The varied activities and programs of NAM did not evolve randomly or by accident. Instead they were the by-product of implementing NAM's By-Laws (revised in 1972, 1974, 1979 and 1994 also, modified in 1997, 1999, and 2019), and NAM's Strategic Plan (1994). In its Strategic Plan, NAM identified operational and programmatic goals associated with its mission and purpose and its functioning as a professional organization. NAM's Five Year Strategic Plan and NAM's collaborations and involvements with other professional organizations in the mathematical sciences have led to its development as a professional organization of quality.

NAM, as a professional organization in the mathematical sciences, has a National Office under the supervision of an Executive Director (formerly an Executive Secretary) with support from the Executive Committee and staff which provides year - round services for its members and the mathematical sciences community.

7. NAM's activities and programs by seasons

Winter. Annual NAM National Meeting which is held in January at the Joint Mathematics Meetings. The standard activities of this meeting include: The

Haynes-Granville-Browne Symposium of presentations by new PhDs American minority mathematicians, the **NAM Panel** which addresses current issues in the mathematical sciences/mathematics education/public policy, the **NAM Business Meeting** for the full membership group, **NAM's Annual Banquet** where the **Cox-Talbot Address** is given as well as various awards and recognitions, and lastly, the **Claytor-Woodard Lecture** is presented.

Spring. NAM's Regional Faculty Conference on Research and Teaching Excellence is designed to encourage more scholarly productivity and improved teaching. It is held on the campus of a HBCU/MSI. The **Bharucha-Reid Lecture** is given by an established mathematical scientist.

Summer. NAM's David Blackwell Lecture, at MAA Summer MathFest provides an opportunity for minority mathematicians to give a scholarly presentation with no major opposing scheduled activity at that time.

Fall. NAM's Undergraduate MATHFest is held annually to encourage students to pursue advanced degrees in mathematics and mathematical education. It is held on the campus of a HBCU/MSI. In addition, an established mathematical scientist is invited to give the **J. Ernest Wilkins Lecture**.

8. Publications

Quarterly newsletters. NAM published its first newsletter in 1971 and has published newsletters throughout its five decades of existence. For the past few decades, NAM has published a quarterly newsletter. In fact, NAM's newsletter has been one of its most effective instruments for communicating with its members and with the larger mathematical-sciences community.

NAM's Proceedings. After the 1980 National Meeting, NAM published its first Proceedings. It was a very important document in that it introduced NAM's annual National Program, including NAM's inaugural Claytor Lecture and Cox-Talbot Address. The editors of this historic document were M. Solveig Espelie, Paul Slepian, and James A. Donaldson of Howard University.

Other NAM Proceedings were published in 1988 and 1989. Several of NAM's Proceedings were planned in the early 1990's but were never published. NAM published a 30th Anniversary Proceeding in 1999 and plans to publish other Proceedings in the future.

Other publications by NAM. NAM has produced several position papers that had a limited distribution.

NAM has produced one book: *Survey of Minority Graduate Students in U.S. Mathematical Sciences Departments* by John W. Alexander and William A. Hawkins [**6**], Co-Project Directors. This volume was a MAA - NAM joint project.

NAM commissioned two books: *The History of NAM, the First Thirty Years, 1969 -1999* by Johnny Houston [**1**], published in 2000; and *Profiles of African American Mathematicians* by Johnny Houston [**3**], to be published in 2020.

9. NAM's governance - organizational structure

NAM's Board of Directors is elected by the General Membership of NAM and it is responsible for the activities, programs and business affairs of the organization. The National Office is managed by the Executive Director. The Corporation (NAM) has six organizational levels (Tiers) at which the affairs of NAM are conducted. The organizational levels are listed, beginning with the Tier directly involving the largest number of persons.

A. Tier I - General Membership
   The general membership shall consist of all the individuals who are currently financial members for the period of time under consideration.

B. Tier Il - Institutional Representatives
   Institutional representatives consist of all those persons selected/appointed by the State/Area Representatives and confirmed by the Board of Directors to serve as NAM's liaison persons at Historically Black Colleges and Universities and Minority Serving Institutions of higher learning (HBCU/MSI) or other institutions of higher learning with significant numbers of NAM members.

C. Tier Ill - State/Area Representatives
   State/Area Representatives shall be persons selected by NAM's Regional Representatives (and confirmed by Board) to be NAM's State/Area liaison persons.

D. Tier IV - Regional and Special Interest Representatives
   The regional and special-interest representatives shall consist of persons elected by the general membership to represent designated geographical regions and designated special Interest groups. Each person duly elected is to serve as NAM's liaison person as well as NAM's Coordinator of Activities for that region/special interest group. These persons are also members of NAM's Board of Directors.

E. Tier V - Board of Directors
   The Board of Directors shall consist of persons elected by the General Membership of NAM to officially manage the affairs of NAM, including a president, a vice president, a secretary, a treasurer, an editor (appointed) and NAM's Regional and Special Interest Representatives. Currently the members are as in Table 2.

10. Some major influences of NAM over the past 50 years

A. **In 1969, NAM was founded.**

B. **In 1971, AWM was founded** partially because many activities and actions of the majority group that excluded AWM's primary membership.

C. **In 1973, SACNAS was founded**, partially because many activities and actions of the majority group that excluded SACNAS's primary membership; SACNAS is the Society of Chicanos/Hispanics and Native Americans in Science.

FIGURE 6. Most of NAM's 2019 Board Members: Ulrica Wilson, Omayra Ortega, Roselyn Williams, Edray Goins, Naiomi Cameron, Johnny Houston, Carla Cotwright-Williams, and Michael Young

TABLE 2. NAM's Board of Directors in 2019

President	Edray Goins	Pomona College
Vice-President	Naomi Cameron	Spelman College
Executive Director	Leona Harris	Univ. of the District of Columbia
Secretary	Shea Burns	NC A & T State University
Treasurer	Cory Colbert	Washington and Lee University
Editor	Omayra Ortega	Somona State University
Region A Member	Chinenye Ofodile	Albany State University
Region B Member	Karen Morgan	Johnson C. Smith Univ.
Region C Member	Brittany Mosby	TN Higher Education Commission
Majority Institution Member	Michael Young	Iowa State University
Outside of Academia	Carla Cotwright-Williams	Department of Defense
Community College Member	Karen Taylor	Bronx Comm. Coll.
NAM Historian	Johnny L. Houston	Elizabeth City State University
President Emeritus	Edray Goins	Pomona College

D. **In 1975, NAM began having its annual national meeting in January** at the JMM (Winter Joint Mathematics Meetings) as a Junior Partner. NAM held its first NAM Panel and in 1975, and NAM appointed Johnny Houston as its Executive Secretary who served until 2000.

E. **In 1976, Howard University established the first mathematics PhD program at an HBCU** under the leadership of James Donaldson, a founder of NAM, and with the support of J. Ernest Wilkins, a great supporter of NAM. This program has produced more African American PhDs in mathematics than any other institution in the world, as of 2019.

F. **In 1977, MAA and NAM, initiated the BAM Program (Blacks in Mathematics) under the leadership of Etta Falconer** (BAM's first national director), an officer on NAM's Board; NAM would provide speakers and MAA would provide financial support for African American mathematicians to speak to students in colleges and high schools to encourage their interest in mathematics; the program lasted at least a decade.

G. **In 1979, NOAA (National Oceanic and Atmospheric Administration) provided NAM with a grant (NAM's first grant)** to have NAM's 10th Anniversary National Conference in Boulder, CO. It supported over 125 faculty members from HBCUs/MSI; NOAA's Environmental Research Labs was located in Boulder and Beauregard Stubblefield, a NAM founder, was working as a senior research mathematician in their labs and helped to get the grant.

H. **In 1980, the book: *Black Mathematicians and Their Works* was published by Virginia Newell**, one of NAM's early newsletter editors, Beauregard Stubblefield, one of NAM's founders, and two of their colleagues; it was the first major publications about African American mathematicians as a significant entity. **NAM published its first proceedings and established the Claytor-Woodard Lecture and the Cox-Talbot Address.**

I. **In 1984, Professor Albert Turner Bharucha-Reid delivered the first NAM Claytor-Woodard Lecture** at NAM's National Meeting in Louisville, KY.

J. **In 1985, Professor David Blackwell delivered the second NAM Claytor-Woodard Lecture** at NAM's National Meeting in Anaheim, CA.

K. **In 1986, Professor J. Ernest Wilkins delivered the third NAM Claytor-Woodard Lecture** at NAM's National Meeting in New Orleans, LA.

L. **In 1988, NAM held a Banquet at its National Meeting in Atlanta, GA in honor of Evelyn Boyd Granville** (who could not attend due to inclement weather) and Marjorie Lee Browne, posthumously. Also, in 1988, **NAM produced its second proceedings.**

M. **In 1989, NAM produced its third Proceedings.**

N. **In 1990, Scott Williams delivered the Claytor-Woodard Lecture and Johnny Houston delivered the Cox-Talbot Address** at NAM's National Meeting in Louisville, KY. During the same year NAM members Johnny Houston, James Donaldson, and Daniel Williams attended ICM-90 in Kyoto, Japan. William Hawkins was named first Director of MAA SUMMA (Strengthening Minority Mathematics Achievement) Office.

O. **In 1992, Gloria Gilmer became the first woman to give the Cox-Talbot Address** at NAM National Meeting in Baltimore, MD.

P. **In 1993, Fern Hunt became the first woman to give the Claytor-Woodard Lecture** at the NAM National Meeting. NAM Undergraduate MATHFest became a regular annual program for NAM. James Donaldson, a founder and former newsletter editor of NAM was elected 2nd vice president of MAA for 1994-96.

Q. **In 1994, during NAM's 25 Anniversary Year, NAM significantly expanded NAM's annual program offerings** and named lectures by establishing:

(a) **NAM Regional Faculty Conference on Research and Teaching Excellence** in the spring and its associated Special Lecture, the **Albert Turner Bharucha-Reid Lecture**,

(b) **NAM-MAA David Blackwell Lecture**, with the honoree invited to give the inaugural lecture,

(c) **NAM Undergraduate MATHFest** in the fall and its associated **J. Ernest Wilkins Lecture**, and

(d) NAM established its **Lifetime Achievement Award** and selected the **first two awardees: David Blackwell and J. Ernest Wilkins.**

R. **In 1995, CAARMS** (Conference of African American Researchers in the Mathematical Sciences) **was established**, hosted at MSRI (June 21 - 23) and organized by Raymond Johnson, William Massey and James C. Turner, members of NAM. The Conference attracted approximately 80 attendees, half of which were African Americans who had earned a PhD in mathematics. During this same year **Lee Lorch received NAM's third Lifetime Achievement Award** and John W. Alexander, president of NAM and Johnny Houston, Executive Secretary of NAM, attended the **Fourth Pan-African Congress of mathematicians in Ifrane, Morocco. SIAM held its first Diversity Day.**

S. **In 1996, William Massey delivered the Claytor-Woodard Lecture** and **Evelyn Boyd Granville delivered the Cox-Talbot address** at NAM's National Meeting in Orlando, FL; **Granville also received NAM's Lifetime Achievement Award**. NAM held its first summer Computational Science Institute. Raymond Johnson completed five years as chair of mathematics at University of Maryland at College Park College Park; during his tenure, a large number of African Americans successful pursued graduate study in mathematics at the University of Maryland and earned a PhD degree.

T. **In 1997, Charles B. Bell delivered the Claytor-Woodard Lecture** and **Carolyn Mahoney delivered the Cox Talbot Address** at NAM's National Meeting in San Diego, CA; **Bell received NAM's Lifetime Achievement Award**. Sylvia Bozeman was elected Governor of the Southern Section of the MAA, She was the first African American to serve as an MAA Governor of a Section; Clarence Stephens was elected in 1962 to serve as Governor of the Maryland Section; because he moved to New York immediately after the election, he never served.

FIGURE 7. Some NAM Members Attending CAARMS at Duke University in 2002: Arlie Petters, Freeman A. Hrabowski III, Johnny Houston, J. Ernest Wilkins, and William A. Massey

The Joint Summer Mathematics Meeting changed to the name: MathFest or MAA MathFest. NSA held its 5th Invitational Conference to attract underrepresented Americans to the agency. NAM and MAA published a book on a Survey of Minority Graduate Students in Mathematics and AMS published CAARMS2 Proceedings; Jack Alexander and Williams Hawkins were authors of the first and Nathaniel Dean was the editor of the latter.

U. **In 1998, Joshua Leslie delivered the Claytor-Woodard Lecture** at NAM's National Meeting in Baltimore, MD. **Clarence F. Stephens received NAM's 6th Lifetime Achievement Award. Mathematicians of the African Diaspora (MAD) webpages [7] were established by Scott Williams at SUNY-Buffalo**, featuring the profiles of all Black mathematicians that could be identified at the doctoral level; it was the most extensive and useful resource about Black Mathematicians since the Virginia Newell et al resource book in 1980 and is still useful today. Johnny Houston and James Turner were invited to present at the Third Edward A. Bouchet Conference that was held in Gaborone, Botswana. Life by Numbers, a seven-part PBS Series that featured William Massey and Nathaniel Dean, revealed the role of mathematics in all aspects of society. It first aired on April 8.

V. **In 1999, Earl Barnes delivered the Claytor-Woodard Lecture** and **Johnny Houston delivered the Cox-Talbot Address** at NAM's National Meeting in San Antonio, TX. Houston announced his retirement as Executive Secretary of NAM, effective June 30, 2000, after 25 years of service. **Houston received NAM's 7th Lifetime Achievement Award** and Leon Woodson was appointed as NAM's next Executive Secretary. Woodson served until 2017.

W. Some Highlights, 2000-2004:

(a) The EDGE (Enhancing Diversity in Graduate Education) Program was established by Sylvia Bozeman of Spelman College and Rhonda Hughes at Bryn-Mar College. The first EDGE summer session was at Bryn Mawr College in 1998 and the location alternated between Bryn Mawr and Spelman Colleges until 2003. Then, it took place at other locations.

(b) **John W. Alexander was President of NAM from 1994-2004.**

(c) **Robert Bozeman was Secretary-Treasurer of NAM from 1991-2004.**

(d) Evelyn Dawley Green, Etta Z. Falconer, and John W. Alexander received NAM's Lifetime Achievement Awards during this period.

(e) The Blackwell-Tapia Lecture Series was established in 2000 to provide a biennial forum for African Americans, Hispanic Americans, and Native Americans to present their research. In 2002 the Blackwell-Tapia Prize was added. Arlie Peters received the 1st prize.

(f) Etta Falconer, a very resourceful member of NAM since 1970, passed in 2002. She was awarded NAM's Lifetime Acheivement Award, posthumously, in 2003.

X. Some Highlights, 2005-2009:

(a) Nathaniel Dean was elected president of NAM; Roselyn Williams was elected Secretary-Treasurer; and Dawn Lott was elected Vice President, all in 2005.

(b) Scott Williams served as NAM's Editor for a decade.

(c) William Massey received the second Blackwell-Tapia Prize.

(d) Sylvia T. Bozeman received AAS Mentor Award in 2009.

(e) The Infinite Possibilities Conference for Women of Color in Mathematics and Statistics (they awarded the Etta Falconer Mentor Prize) was held at NC State University in 2008 and at other universities during other years.

(f) Many summer research programs for American minorities and women were held at several universities. The one at Miami University of Ohio, SUMSRI, had been operating for a decade (1999-2009).

(g) An extraordinary turn of events occurred during the Fall of 2009, when Dr. Raymond Johnson stepped to the front of a Rice University classroom for the first time. Johnson, the first black student to earn a degree at Rice University in 1963, had returned as a full professor.

Y. Some Highlights, 2010-2014:

(a) two mathematical giants passed during this period: David Blackwell (2010) and J. Ernest Wilkins (2011).

(b) All the other many programs continued; a woman won the Blackwell-Tapia Prize, Trachette Jackson; NAM's Lifetime Award was received by Abdulalim Shabazz (2010), Robert Bozeman (2011), Sylvia Bozeman (2012) and William Hawkins (2014).

(c) There are eight NSF funded Mathematical Sciences Institutes that have a wide range of exciting, cutting-edge programs. The Institutes formed a Diversity Committee to insure the participation of underrepresented American minorities and women. They are: 1. the American Institute of Mathematics (AIM, `http://www.aimath.org`); 2. the Institute for Advanced Study (IAS, `http://www.math.ias.edu`); 3. the Institute for Computational and Experimental Research in Mathematics (ICERM, `http://icerm.brown.edu/`); 4. the Institute for Mathematics and its Applications (IMA, `http://www.ima.umn.edu`); 5. the Institute for Pure and Applied Mathematics (IPAM, `http://www.ipam.ucla.edu`); 6. the Mathematical Biosciences Institute (MBI, `http://www.mbi.osu.edu`); 7. the Mathematical Sciences Research Institute (MSRI, `http://www.msri.org`); and 8. the Statistical and Applied Mathematical Sciences Institute (SAMSI, `http://www.samsi.info`).

Z. Some Highlights, 2015-2019:

(a) Edray Goins became President of NAM and Talitha Washington became Vice President, both occurring in 2015, and Washington continued as NAM's editor.

(b) African Americans earned only 1.4 percent of all mathematics doctorates in 2015.

(c) Professor Abdulalim Shabazz passed (2014) and NAM had a Special Session in his honor in 2015 at MAA MathFest.

(d) The Department of Mathematics at Morehouse College was chosen to receive the 2016 AMS Mathematics Programs that Make a Difference Award, the first HBCU to receive the award.

(e) At the JMM in 2017, NAM held a Special Session on "Mathematics in the Atlanta University Center."

FIGURE 8. Some Blackwell-Tapia Prize Winners and NAM Members from November 2018: William A. Massey, Michael Young, Isom Herron (back), Ulrica Wilson, Nathaniel Whitaker, Edray Goins (front), Asamoah Nkwanta, Talitha Washington, Donald King (back), Arlie Petters, Ronald Mickens, Johnny Houston, Ebony Harvey, and Mel Currie

(f) The movie "Hidden Figures" was released in December 2016 and Morehouse College professor Rudy Horne was the mathematical consultant. Mathematician Katherine G. Johnson became a household name. NAM participated in a Hidden Figures Session at the JMM in 2017.

(g) AWM and Spelman continued the Etta Falconer Lectures each year.

(h) The MAA Alder Award is a distinguished teaching award that recognizes – at most – three, beginning college or university math faculty members each year, whose effectiveness in teaching undergraduate mathematics illustrates influence beyond individual classrooms.

(i) Project NExT is a valuable experience for new mathematics faculty; both the Alder Award and Project NExT have selected a number of African American participants, recently.

(j) NAM presented Centenarian Awards to Virginia Newell (2018), Clarence Stephens (2018), and Catherine Johnson (2019). Lifetime Achievement Awards were presented to Jacquelyn Giles (2015), William Hawkins (2016), Nagambal Shaw (2017), Rudy Horne (2018), Carolyn Mahoney (2018), and Melvin Currie (2019).

(k) NAM launched a 2-million-dollar Endowment Campaign which will continue to the end of 2019.

(l) At the 2019 JMM in Baltimore, NAM began a year-long celebration of its 50th Anniversary Year, honoring its founders, presenting Evelyn Boyd Granville with a Legacy Award and having a Special Session on NAM at the JMM.

11. Conclusion

It was interesting to note that the first African American to earn a PhD in mathematics, Elbert F. Cox, died the year (1969) that NAM as founded. It is also interesting to note that there were only sixty-nine (69) African Americans who had earned a PhD in mathematics before the Founding of NAM. That was an average of about 2 PhD's per year since Cox earned the first such degree.

Fifty years later, more than 400 additional African Americans have earned a PhD in the mathematical sciences, an average of 8 PhD's per year. Although these numbers and the events listed above reflect the progress that has been made thus far, there is much left to be desired. The average number of African Americans receiving PhD's in the mathematical sciences each year is still around 1% or less. The progress that has been made is far short of NAM's (Visionary Goal – G). Much more effort will be needed. To meet these challenges is NAM's focus for the next 50 years.

NAM's ORIGIN WAS DRAMATIC, ITS GOALS ARE NOBLE, ITS JOURNEY HAS BEEN IMPACTFUL AND ITS FUTURE IS VERY PROMISING. HAPPY GOLDEN ANNIVERSARY YEAR, NAM!

Photo Credits

All photographs in this article are from the personal library of Dr. Johnny L. Houston.

References

[1] Johnny L. Houston, *A brief history of the National Association of Mathematicians, Inc*, African Americans in mathematics, II (Houston, TX, 1998), Contemp. Math., vol. 252, Amer. Math. Soc., Providence, RI, 1999, pp. 139–164, DOI 10.1090/conm/252/1749769. MR1749769

[2] Johnny L. Houston. Private Correspondences and Photographic Collection.

[3] Johnny L. Houston. *Profiles of African American Mathematicians.* NAM (2020).

[4] Johnny L. Houston, *Ten African American pioneers and mathematicians who inspired me*, Notices Amer. Math. Soc. **65** (2018), no. 2, 139–143, DOI 10.1090/noti1639. MR3751310

[5] Johnny L. Houston, *A centennial year (2019) reflection on the life and contributions of mathematician David H. Blackwell (1919–2010)*, Notices Amer. Math. Soc. **66** (2019), no. 2, 221–226. MR3840124

[6] John W. Alexander and William A. Hawkins. *Survey of Minority Graduate Students in U.S. Mathematical Sciences Departments.*

[7] Mathematicians of the African Diaspora. `http://www.math.buffalo.edu/mad`.

[8] NAM Newsletter Archive. `https://www.nam-math.org/archives.html`.

602 West Main Street, Elizabeth City, North Carolina 27909
Email address: `jlhouston602@gmail.com`

Selected Published Titles in This Series

759 **Omayra Ortega, Emille Davie Lawrence, and Edray Herber Goins, Editors,** The Golden Anniversary Celebration of the National Association of Mathematicians, 2020

756 **Joeri Van der Veken, Alfonso Carriazo, Ivko Dimitrić, Yun Myung Oh, Bogdan D. Suceavă, and Luc Vrancken, Editors,** Geometry of Submanifolds, 2020

755 **Marion Scheepers and Ondřej Zindulka, Editors,** Centenary of the Borel Conjecture, 2020

754 **Susanne C. Brenner, Igor Shparlinski, Chi-Wang Shu, and Daniel B. Szyld, Editors,** 75 Years of Mathematics of Computation, 2020

753 **Matthew Krauel, Michael Tuite, and Gaywalee Yamskulna, Editors,** Vertex Operator Algebras, Number Theory and Related Topics, 2020

752 **Samuel Coskey and Grigor Sargsyan, Editors,** Trends in Set Theory, 2020

751 **Ashish K. Srivastava, André Leroy, Ivo Herzog, and Pedro A. Guil Asensio, Editors,** Categorical, Homological and Combinatorial Methods in Algebra, 2020

750 **A. Bourhim, J. Mashreghi, L. Oubbi, and Z. Abdelali, Editors,** Linear and Multilinear Algebra and Function Spaces, 2020

749 **Guillermo Cortiñas and Charles A. Weibel, Editors,** K-theory in Algebra, Analysis and Topology, 2020

748 **Donatella Danielli and Irina Mitrea, Editors,** Advances in Harmonic Analysis and Partial Differential Equations, 2020

747 **Paul Bruillard, Carlos Ortiz Marrero, and Julia Plavnik, Editors,** Topological Phases of Matter and Quantum Computation, 2020

746 **Erica Flapan and Helen Wong, Editors,** Topology and Geometry of Biopolymers, 2020

745 **Federico Binda, Marc Levine, Manh Toan Nguyen, and Oliver Röndigs, Editors,** Motivic Homotopy Theory and Refined Enumerative Geometry, 2020

744 **Pieter Moree, Anke Pohl, L'ubomír Snoha, and Tom Ward, Editors,** Dynamics: Topology and Numbers, 2020

743 **H. Garth Dales, Dmitry Khavinson, and Javad Mashreghi, Editors,** Complex Analysis and Spectral Theory, 2020

742 **Francisco-Jesús Castro-Jiménez, David Bradley Massey, Bernard Teissier, and Meral Tosun, Editors,** A Panorama of Singularities, 2020

741 **Houssam Abdul-Rahman, Robert Sims, and Amanda Young, Editors,** Analytic Trends in Mathematical Physics, 2020

740 **Alina Bucur and David Zureick-Brown, Editors,** Analytic Methods in Arithmetic Geometry, 2019

739 **Yaiza Canzani, Linan Chen, and Dmitry Jakobson, Editors,** Probabilistic Methods in Geometry, Topology and Spectral Theory, 2019

738 **Shrikrishna G. Dani, Surender K. Jain, Jugal K. Verma, and Meenakshi P. Wasadikar, Editors,** Contributions in Algebra and Algebraic Geometry, 2019

737 **Fernanda Botelho, Editor,** Recent Trends in Operator Theory and Applications, 2019

736 **Jane Hawkins, Rachel L. Rossetti, and Jim Wiseman, Editors,** Dynamical Systems and Random Processes, 2019

735 **Yanir A. Rubinstein and Bernard Shiffman, Editors,** Advances in Complex Geometry, 2019

734 **Peter Kuchment and Evgeny Semenov, Editors,** Differential Equations, Mathematical Physics, and Applications, 2019

733 **Peter Kuchment and Evgeny Semenov, Editors,** Functional Analysis and Geometry, 2019

For a complete list of titles in this series, visit the
AMS Bookstore at **www.ams.org/bookstore/conmseries/**.